KT-379-500

Last of the Line

Traditional British Craftsmen

Tom Quinn & Paul Felix

DAVID & CHARLES

Contents

81%
FLOUR
WATERMILL
MILLSTONE GROUND
FLOUR

Introduction

The twentieth century has seen more changes in the way we live than any previous single century. In 1900 the motorcar was still extremely rare; man had not flown; the pace of life throughout the world was still dictated by the horse, and much that was used in everyday life was still created by hand.

Now as we reach the end of the century we have tapped the vast resources of nuclear power; we have explored the outer reaches of space; we can fly across the world in enormous comfort in less than a day, and we routinely employ computers and telecommunication systems of staggering complexity. But in the midst of these momentous technological changes, as we reach the beginning of the second millennium it is remarkable how many ancient crafts that were already in decline at the beginning of the twentieth century still survive.

It is true that many survive in the hands of just one or two individuals working quietly away in often remote bits of rural Britain, but the fact that they are still with us at all is cause for celebration for they are part of our history, part of a past of craftsmanship, skill and attention to detail that most of us probably thought had vanished forever.

The purpose of this book is to record and celebrate the best of these ancient crafts before they disappear – and, more importantly, to record the lives of the craftsmen and women themselves.

Men like Gerald Haynes, one of the last remaining freeminers in the Forest of Dean, or Eustace Rogers, who made coracles on the Severn at Ironbridge as they were made by his ancestors three centuries ago; men like Jack Durden who still makes charcoal deep in a Buckinghamshire woodland in a tradition dating back to Roman times.

And then there are the trug makers, the flint knappers, the reed cutters, the rake and hurdle makers – all are still with us.

In this book we have tried, in words and pictures, to capture the essence of what these ancient craftsmen do; to record a world whose origins lie in the heart of pre-industrial Britain.

The Cowskin Boatman

In a report commissioned by the Campaign for the Protection of Rural England, Shropshire was listed as one of only a handful of English counties which still retain their ancient rural aspect. For in Shropshire one can still walk for miles without coming across a sign of human habitation, and the pattern of woodland, field and hedgerow has been left relatively unscathed by the worst excesses of modern industrial agriculture. If the countryside of Shropshire changes slowly, so too do the towns and villages, for this is a land resistant to the wiles of temporary fashion.

Among Shropshire's most famous towns is Ironbridge. And here, soaring over the Severn Gorge, is the world's oldest cast-iron bridge – the bridge that gives the town its name. Built in 1779 by Abraham Darby (1750–1791), whose grandfather built the first coke-fired blast furnace, the bridge remains pretty much unchanged to this day.

Tucked away beneath the great arch of the bridge in a little house and workshop that is famously difficult to find, has no telephone and seems older even than the bridge, one might – until very recently – have been lucky enough to find Eustace Rogers, the last of the traditional coracle makers.

I say 'the last' despite the fact that one or two others also make these ancient leathern boats, because Eustace, now in his eighties, has coracle making in his blood. Where other makers have found an interesting hobby in coracle building and have perhaps come to it late in life, Eustace has been a coracle maker all his life; he learned his skills as a boy from

Opposite: How to carry your coracle. Eustace Rogers shows how it's done

Left: Tommy Rogers, celebrated Ironbridge poacher and coracle-maker, whose exploits have been recorded for posterity via interviews with his 83-year-old grandson Eustace Rogers, the last coracle-maker in the Gorge

Opposite: There are no special tools needed for coracle making – just a hammer, nails, a saw and a chisel

Left: The waterproof underside of the cowskin boat

his father, and through father and son the family connection with coracles goes back through several centuries. In fact the Rogers family were certainly making coracles here long before the iron bridge itself was built.

We know relatively little about the origins of the coracle, but the ancient Celts certainly used them to catch the salmon that came up the Severn each year in their tens of thousands, before pollution and high seas netting effectively destroyed the king of fishes.

Two thousand years ago Julius Caesar and Pliny mentioned coracles in their writings, as did diarist John Aubrey (1626–1697) in the late seventeenth century. They were certainly still widely used on the Severn and elsewhere in the Celtic world throughout the eighteenth and nineteenth centuries, and the decline in their use really only started in the later decades of the nineteenth century. Even after the First World War they were still to be seen along the Severn, which was their last stronghold, but in 1920 a Salmon and Fisheries Act put an end to coracle fishing on most rivers.

Eustace Rogers only recently made his last coracle, and he only gave up when failing health forced him to it. A small, bright-eyed man who loves to chat and takes a lively interest in everything from local politics to the state of the river, he has the gnarled yet expressive hands of the true craftsman. And though he no longer makes coracles, his memories of coracle making with his father and grandfather are as fresh as ever.

'When I was a boy the river was full of the small round boats, the coracles,' he says, 'and the river was full of fish, too. You could scoop them out easily at certain times of year. But the coracles took some skill to use – nothing like an ordinary boat. If you don't know what you're doing you never get anywhere in them – you'll just go round in circles.'

Eustace laughs, and then goes on to explain that his grandfather belonged to a gang of poachers who would paddle up and down the river at night.

'Coracles were wonderful for poaching,' he says with a grin, 'because once you'd caught your fish you were out of the water with the boat on your back and gone into the woods in a second.'

Eustace was born in Ironbridge, and he was twelve when his father set him to 'rearing' coracles. For the next seventy years he did nothing else. Working in the old family home under the bridge he discovered the ancient techniques for bending and shaping willow and hazel for the frame and basket. The basket was then covered with a whole stretched cowskin.

'Some modern ones are covered with calico, which is lighter,' says Eustace, 'but a cowskin boat will last much longer and it was a design that had probably remained unchanged in more than a thousand years.'

Eustace's last coracles were sold mainly to museums and a few anglers.

'A coracle has no keel or rudder, so to paddle it you need to learn what we might best describe as a long figure-of-eight movement over the front of the boat.'

All the wood used for the basket frame came from local sources – even the wide-bladed paddle was made by Eustace to a unique design and from local wood. Ironbridge coracles measure roughly 5ft in diameter, and to the expert eye they can be distinguished easily from all other coracles, whether those made on Welsh rivers or even from other places along the Severn.

Eustace Rogers never married and has no children so, sadly, the family connection with coracle making will die with him; but he has trained at least one other man who will carry on this extraordinary craft.

Opposite: Eustace afloat under the famous landmark of Ironbridge

The Bark Tanner

Preparation is vitally important: Molly carefully removes every scrap of fat from the untreated hide (above)

Opposite: Molly stretches a traditonally tanned sheepskin. Bark tanning may be slow and old-fashioned, but results in a vastly superior product

Tanning sheepskins in the traditional way has been Molly Arthur's occupation for over thirty-five years. When she started the business she and her husband, Andrew, were living in a remote part of Inverness.

'We kept sheep for their meat, and it seemed a pity to throw the skins away,' says Molly, 'so I taught myself to cure them. I was lucky to acquire an old book with ancient recipes on tanning, and with that I started to experiment. It was then that I also realised that bark tanning was becoming a lost art, despite the fact that it produces the very best leather and, as regards my trade, the softest, most luxurious sheepskin.

'Most modern tanning involves chemicals that turn the skins a horrible grey. However, I use only bark for my skins. I tried oak bark, which is the oldest way, but it can take up to eight months to cure a skin with oak, so I've settled on powdered bark from African mimosa trees; I buy this by the sackload from a Midlands supplier, and it's terribly expensive! It's a bit of a compromise, but you get the same effect that you'd get with oak bark, but in far less time.'

A small, but robust and genial woman, Molly bustles around her workshop with the enthusiasm of a woman half her age. She clearly enjoys her work, but particularly the sense that she is helping to keep alive a skilled and highly individual trade – a trade moreover that avoids chemicals and produces something lasting, attractive to look at and useful. She worries a little that mimosa is not quite up to the standard of oak, but considers that it is a good and reasonable compromise: 'A skin tanned with mimosa takes about four weeks and gives very good results,' she says.

When the skins first arrive at the Arthurs' home, which is now on the Mull of Kintyre, Molly gets to work with a knife to make sure any remnants of fat have been removed; this is vitally important if the tanning process is to work well, as she explains: 'The skins have to be really clean before I put them in my liquor. I scrape off every last bit of extraneous material – the more work you do at this early stage, the better the final product. The liquor we use is a mixture of water, salt and mimosa bark, and during its four weeks in the mixture each skin is moved about by hand every day. But what, you ask, does this curious mixture do to the sheepskin?

'Well, it changes the whole structure of the skin. You're actually changing the molecules in order to transform skin, which would dry out and crack and crumble if left untreated, into leather which will last for decades – and it's only when you see my bark-tanned skins that you realise how different they are from chemical skins. The bark doesn't dye the wool, and the leather holds the wool tight. It even smells different – just glorious leather and wool.'

The Bee-skep Man

Farmer David Chubb started to make traditional bee-skeps nearly twenty years ago when he decided he wanted one and discovered that no one was making them.

'I'd always kept bees, and skeps were the traditional way to keep them – none of these modern wooden hives. I had no idea what to do once I discovered I couldn't buy a skep, but then I had two strokes of luck. First I found a book describing bee-skeps and how they were made, and then I discovered someone locally who could speak Dutch, and the two went together very nicely as the bee-skep book was in Dutch!'

Even after this lucky break it still took David over a year to gather together all the materials he needed: 'I eventually discovered that wheat straw was the best straw for a skep – I got it locally from an agricultural research station after meeting a man who made corn dollies. So as you can see, it was all chance meetings and luck; but I was very determined!'

David explains that for centuries skeps were made using bramble and hazel to keep the straw together, but his one concession to the modern world is to use cane instead, the kind used by chair makers. The technique for building up the woven straw is very similar to that used for straw mats:

'Yes, it's really quite straightforward, although difficult to describe exactly. You sort of weave in towards the centre

Opposite: David Chubb with a storeroom full of skeps: despite their old-fashioned appearance, these traditional bee hives have many advantages over their modern counterparts. They provide superb insulation, for example, and properly looked after may last a century or more

Left: Bee-skep hives of precisely this type have probably been in existence for as long as man has kept bees. When in use the skeps are given a waterproof coating of wax by the bees themselves

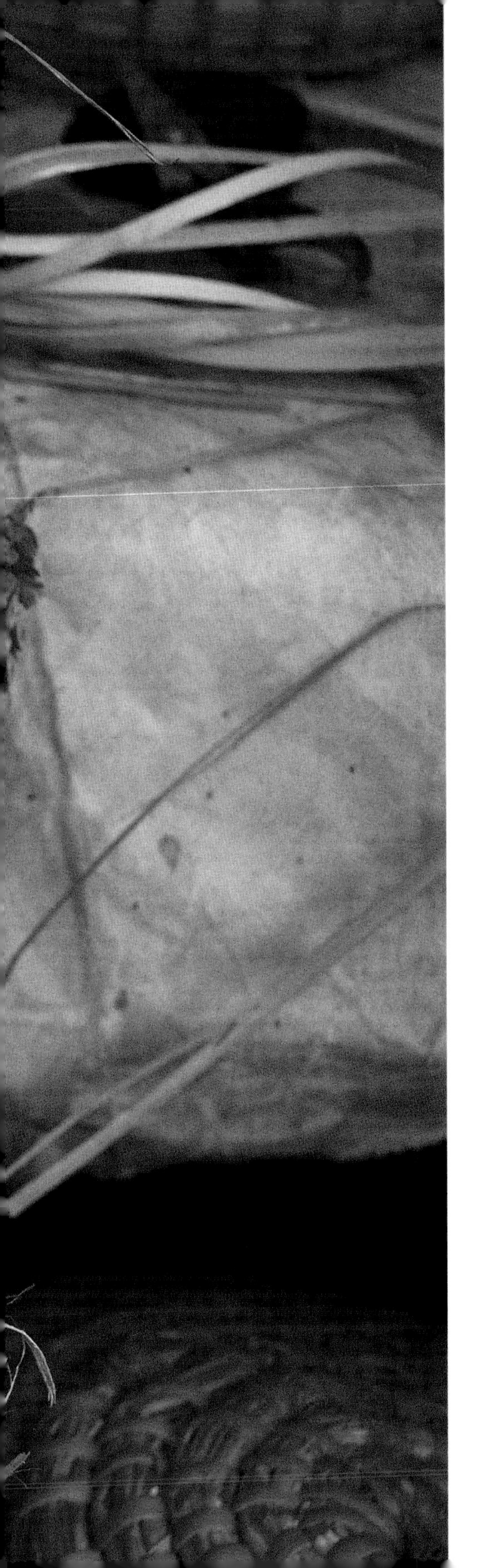

– as in the table mat – then weave the straw out again and back into itself, and gradually the whole thing builds up. As I'm also a farmer I tend to do a couple of hours in the morning and then again in the evening. It's only ever really been a side-line, because as you can probably imagine, the demand for this sort of traditional beehive is very limited. They are more popular in Holland. In my best year – and I've been making them for nearly twenty years, remember! – I made something like 230 for the Chelsea Flower Show. I once made one that was 4½ft high and 4ft wide, but the normal size is 14in wide and 9in high.'

The dense 'tube' of wheatstraw is tied using a long coiled strip of cane, to produce a skep that hasn't altered in design for centuries

A bee-skep is just a hollow basket without all the extra internal arrangements that a modern wooden hive has. 'Modern wooden hives are just designed to get more money out of beekeepers,' says David. 'OK, it is a little easier to check inside a modern hive, but the bees are always just as happy – I think probably happier – in straw. After all, it takes 6in of wood – thickness, I mean – to produce the insulating power of 1in of straw.

'Traditionally a bee-skep would have what's called a hackle on top – a sort of straw hat – to keep the weather off, and it would rest on a stone slab. Under these circumstances a skep might easily last a century or more; I know that in Holland they've used the same skeps for upwards of 140 years. This is partly because the bees line the inside of the skep with a protective jelly that makes it waterproof and preserves the straw beautifully.'

These days David uses special long-stemmed wheat straw, some of which comes from the Prince of Wales' farm at Highgrove. Though orders have declined, they nevertheless still come in steadily, and David believes that the increasing desire among customers for a natural product produced in a natural environment can only be good news for the bee-skep maker.

545
543
361
SIZE 13
Size 14½
WOODSCREWS
BRASS
SHOE REPAIRS
HSE
Maribelle
MELTONIAN SHOE CREAM
NEUTRAL 1
550/327
COGNAC 72
550/340
GREY
KIWI
SHOE POLISH
BRIWA

The Boot-tree Man

A lot of people have no idea that anyone is still making boot-trees,' says Bill Bird. 'They may have ridden for years and they may even own an old pair of boot-trees, but in most cases when the trees get broken or badly worn they just throw them in a corner and forget them. If they brought them to me they could be repaired.'

But why would anyone want a boot-tree? Bill is a passionate advocate of them, as you would imagine: 'You have to remember that a pair of riding boots is really just a piece of leather folded in half and then attached to a shoe, and if you don't keep them in properly made trees they quickly sag, bulge and generally look awful. A boot-tree is, if you like, a piece of furniture that re-moulds the boot every time it is worn.'

A number of companies mass-produce boot-trees, but Bill is one of a tiny number – probably not amounting to more than two or three – who still make them by hand, and to fit individual boots: 'Most boot-trees are simply made on a lathe – it takes about ten hours to make a pair – but we look at each boot and carve the tree to fit exactly. We start with a rough bit of beech and make the basic shape using a bandsaw. After that I use rasps and a sander to get the final shape.'

Beech is the preferred wood as it is fine-grained, hard and not too expensive; says Bill, 'It's also good because it's available in big pieces. Oak is not as good because although it's hard, it is too open-grained; and it can go blue as it is continually wet and then dry as it goes in and out of a damp boot. I used to use tulip wood from the USA, but it's got too expensive.'

Based in deepest Gloucestershire, Bill is able to get good local supplies of timber, but he has also used German and even Romanian beech: 'We don't need a huge amount as I only make twenty to thirty pairs of boot-trees a year, but I also make riding boots – that's

Opposite: Precision is everything in boot-tree making: here Bill Bird carefully measures a section of a foot piece before making a few final adjustments

454
M. Whittaker
493
494
499
499
498
487
487
450
164
542
543
NEUTRAL
550/327
550/41
COGNAC
Shortbread

now the main business, in fact – and we make trees for those, too.

'Some of the work is repair work. People tend to treat their boot-trees rather roughly – they're thrown into the tack room or the back of the Land Rover where they get buried or trodden on, and as they have eight parts it's easy for a bit to get smashed or knocked off or lost. So we also offer a repair service.

'The basic structure of a boot-tree is unvarying. First there is the foot piece, and the shin piece hinges on to this; at the back there is a bulging piece of wood that fills out the calf, and the "key" is a strip of wood that is pushed down the middle of the boot between the calf and the shin pieces – a well-made boot-tree when pushed into its boot will take up all the space and give the boot something to push against, as it were.'

Bill has been making boot-trees for more than twenty years. He learned the trade at a London bootmaker's, but unlike most apprentices who learn first to make boots, Bill was first and foremost a boot-tree maker. 'It took me five years to get a real feel for the business,' he confesses; but these days he's as happy to make spare parts for boot-trees as to make new ones: 'I like the idea of keeping the old ones going because with care there's no reason why they shouldn't last several centuries.'

Left: Bill in his workshop: a century ago everyone who owned a pair of riding boots would also have owned a pair of boot-trees. Today, the art of boot-tree making is fast disappearing

The Bow Maker

Left: *Christopher Boyton tests one of his own hand-made bows – in essentials a weapon that would be instantly familiar to our medieval ancestors*

Despite the widespread use of modern fibreglass bows with fancy sighting arrangements, the traditional English longbow – the deadly weapon that terrified the French at Agincourt – is alive and well. From a small workshop in London, Christopher Boyton, one of the few remaining bowyers in the country, sends his hand-made traditional bows all over the world.

'The best are made from one solid piece of yew 6ft long and tipped with horn to prevent the string cutting into the wood,' says Christopher. 'Bows like this would be instantly recognisable to a medieval archer.'

It takes Christopher just four hours to get the basic shape of the bow, and two full days to complete it. The bowyer's tools couldn't be simpler: traditionally most of the work is done with rasps. The yew arrives sawn square or cleft and about 1¼in square. Yew for bows can come either from the trunk of the tree or from a branch.

'We usually use a bandsaw these days,' says Christopher, 'when removing bow staves from a log as opposed to cleaving, as it is less wasteful of the precious wood. Yew is now rare and expensive and most of it comes from America. I make both laminated and solid yew bows. Although the name does sound rather modern, laminated bows have in fact been made in Europe at least since the sixteenth century, and were made using water-based animal glues, like hide size, but the principle was the same.

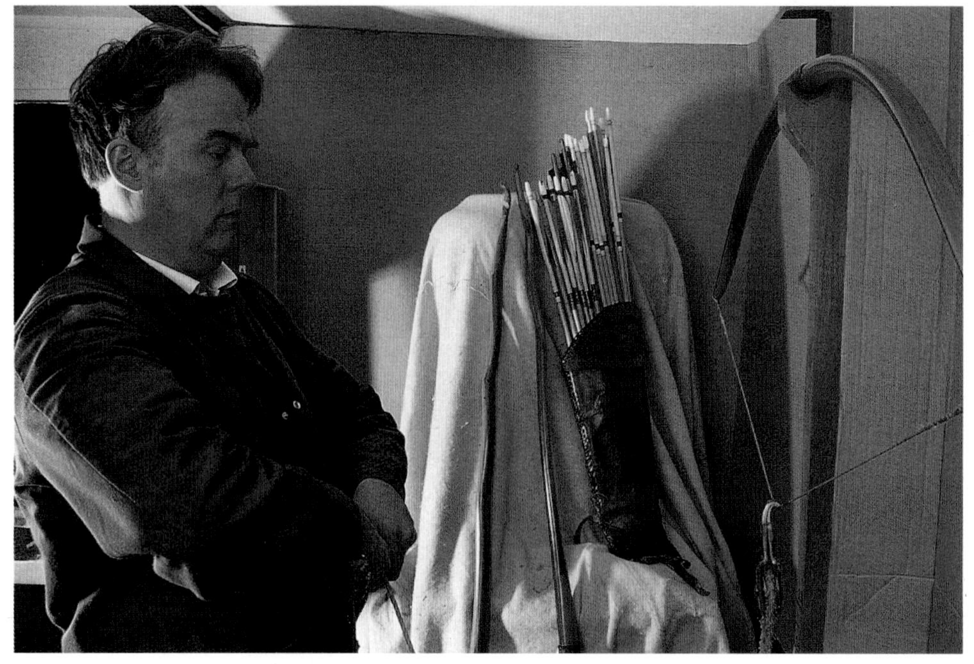

Most laminated bows are made from two pieces of wood, but sometimes three or even four are used.'

With longbows it is the final process that takes the time because the bow is polished till it shines like ivory. The cost of the yew usually amounts to about half the total price of the bow; laminated bows cost a little less than the one-piece of 'self'-yew sort. Christopher also makes arrows for his bows:

'I make the shafts for the arrows from deal, although this is not the wood that would have been used in medieval times. Today we want to use our arrows time and time again, and deal is tough and lasts a long while. In the Middle Ages alder, poplar and aspen were used, quicker-growing,, cheaper woods, because in medieval warfare arrows only needed to be used once – after all, you'd hardly trot over to the enemy to get them back!

'Medieval arrows were slightly longer, at 30in, than their modern counterparts which are 28in – this is judging by those found on the salvaged Tudor warship the *Mary Rose*. Extra length means extra draw and therefore more power. Modern bows draw at about 50lb, but medieval bows were probably 80lb or more. Traditionally the bowstring was made from hemp strands carefully twisted together with some sort of glue, but this art is now completely lost, except in Japan where there is an unbroken tradition of bowstring making.

'The late Richard Galloway, a great bowmaker in Scotland, was the latest in an unbroken line of master bowmakers which stretches back over 300 years; he taught me the basic skills. I can make two to three bows a week, but medieval makers were quicker – but with war in mind they probably had to be.

'I like to make only to order, as each bow is tailor-made for the individual archer. It has to be a good fit and hold. Most traditional longbows are now used for archery meetings. The range of a bow can vary from 150 to well over 300 yards.'

Christopher started making longbows in 1974 in his spare time, and his business grew to a full-time occupation in 1993. Unlike many bowmakers he began constructing bows before taking up archery. 'To begin with I started making them, first for myself and then for a few members of the local archery club, none of whom used longbows at the time, but who began to order them from me.' Now the main thrust of Christopher's business is making arrow shafts. He still builds bows for historical reconstructions and has been involved in projects for several royal armouries, including lecturing at the Tower of London.

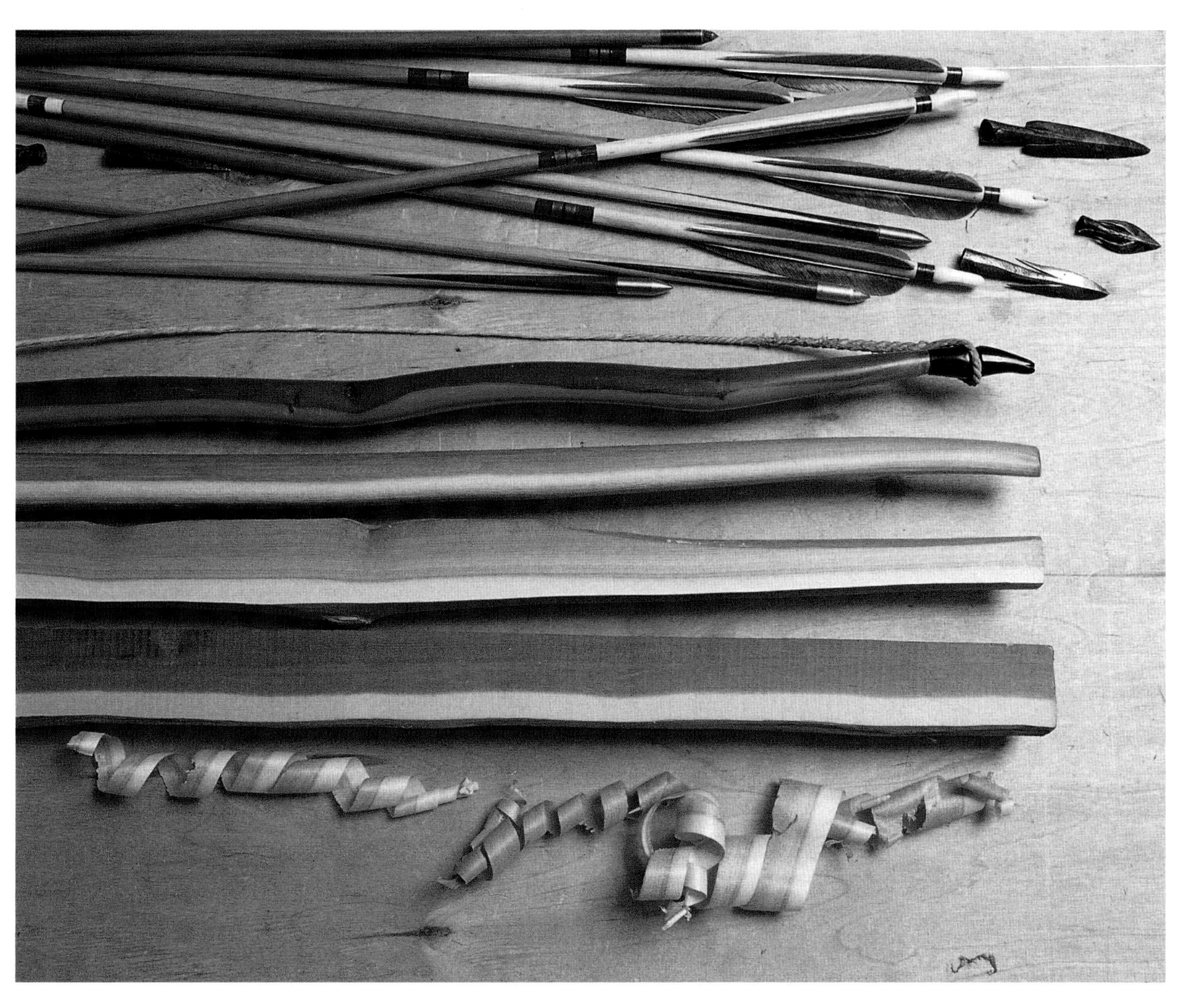

Left: Taking shape: the rough-sawn timber (bottom) is shaped and smoothed until it takes on the traditional longbow shape. The horn tip is fitted to prevent the drawstring cutting into the timber

Opposite top: Precision made: Christopher uses a few simple tools to produce both laminated and 'self' – or single timber – bows. Most go, not to museums, but to individuals who use them regularly

Opposite bottom: Testing the 'draw': modern bows 'draw' at about 50lb, but a medieval bow might be 80lb or more

he Merlin

The Cane Rod Builder

Fishing with rod and line has probably been practised somewhere in the world since time immemorial; its popularity in Britain can be judged by the fact that Isaac Walton's 1653 classic *The Compleat Angler* is one of the most reprinted books in history. Of course, Walton's angler was very different from his modern counterpart, and for the most part would have used a switch of hazel cut from a hedgerow for a rod, with a few lengths of horsehair knotted together and then tied at the end of the hazel. Shop-bought rods and reels – known then as 'winches' – were available, but probably only to a few wealthy individuals.

From the seventeenth to the nineteenth centuries angling grew ever more popular, and anglers and tackle manufacturers searched the world for new and better rod-making materials. The first big breakthrough came in the nineteenth century when greenheart began to be used for rod-making: this springy, supple wood was also immensely strong and could be used to tackle the biggest salmon. It did, however, have one disadvantage, a tendency to dry out and snap without warning.

After greenheart there came what many people still consider to be the ultimate rod-making material: tonkin cane, or to give it its common name, bamboo. Obviously bamboo can be used hollow – in other words, just as it grows – but a technique was pioneered in Britain and America for cutting it into strips that were triangular in section. These tapering strips were then glued together to form a complete rod that would be hexagonal in section.

Despite all the advances of modern materials – most notably, so far as rod building is concerned, carbon and boron – a great number of anglers, particularly trout fishermen, still use cane and argue that it has never been surpassed. Of course, the demand for cane rods is tiny compared to the mass market for carbon and boron rods, but one or two craftsmen, working in out-of-the-way places, still build rods from cane. Shaun Linsley from Stour Provost in Dorset is one of a tiny band of craftsmen who continue to produce these exquisite rods. Each Linsley rod will take eight weeks to make, and as Shaun is the first to admit, the manufacturing process is at times intensely fiddly and intricate. So how is a cane rod made? Shaun is only too happy to explain:

'First we get the best tonkin cane available, and that grows only along the Sui river, about one hundred miles north-west of Canton. It has the perfect combination of strength and power – it's so good in fact that it's used in China to make scaffolding for work on high-rise buildings! It is expensive, however.

'We season our cane for ten years, and this is absolutely essential if the rod is to be as near perfect as anything manufactured can be. We then split the cane into strips:

Opposite: From raw material to finished product: the cane used for Shaun Linsley's rods grows in only one place in the world

Above: Having been straightened using heat and a vice, each strip of cane is planed by hand, a job requiring enormous skill

Left: Shaun checks a batch of cane. Despite space-age materials such as carbon and boron, many anglers believe that as a fishing-rod material cane has never been surpassed

we use the traditional six section – that is, hexagonal – construction method because although other methods involving four and five strips have been tried, they've never really matched the six-strip method. I split the cane by hand rather than cutting it along its length; this adds strength because it follows the line of the fibres rather than cutting through them.

'The next stage requires great precision, when each strip of bamboo has to be made perfectly straight before being triangulated and then tapered. The straightening process is done by hand using heat and a vice, and it is highly skilled work as too much bending and heating will weaken the cane. Each strip is then milled, and planed by hand, and they are all then hardened by being carefully baked in an oven, the idea being to dry out any moisture in the cane and to set it. The six strips are glued together and left for a month; then the rod is sanded to produce a silky finish. A top-quality cork handle is fitted before the reel seat fittings are added; the rod rings are whipped on using the finest silk.'

To the uninitiated this may seem an extraordinarily laboured business – and all to produce a fishing rod. But that's not how Shaun sees it: 'Well, casting thirty feet of line accurately demands a rod made to these standards, and fishermen are a fanatical bunch and more than happy to pay whatever is necessary for a rod that looks and feels just right. You have to remember that a good cane rod makes fishing far more than a mere means to catch fish – with a beautiful cane rod in your hands even *not* catching fish is a pleasure!'

Shaun makes rods only to individual order, but he is as happy to make a 12ft salmon fly rod as a tiny toothpick of a 7ft brook rod. But his reputation has spread, and he now has a waiting list of over a year.

The Cheese Maker

Charles Martell is one of only two people still making single Gloucester farm cheese. He lives in a remote corner of north-west Gloucestershire in the heart of the area in which single Gloucester was once a staple part of the poorer people's diet. Always the poor relation of the better-known double Gloucester, the single variety is now highly prized by cheese connoisseurs – perhaps because of its rarity, perhaps because it is simply very good.

Charles Martell, now in his late fifties, has been making single Gloucester cheese for more than twenty-five years, and he is delighted that under a European Union rule it cannot now be made outside the region where it was first developed. Similarly the EU law prevents champagne being made in Kent, or Greek feta cheese in Yorkshire – 'and a jolly good thing too!' says Charles.

'Traditionally that is the way it always was,' he continues, 'because single Gloucester was never transported and sold, simply because in those days when there were no refrigerated lorries, if you put it on a cart or on a barge it fell apart with all the rattling and moving. So single Gloucester was always and only eaten locally.'

Charles is a real enthusiast as well as a man who believes passionately that farm cheese makers like himself should be able to make a living. That's why when he was approached by a big supermarket chain he told them he wouldn't sell his cheese to them: 'When the big supermarkets get hold of you the next thing that happens is that they squeeze you financially, and what starts as a squeeze ends as financial strangulation. It's already happened with livestock farmers, and I don't want to be next.'

Charles was a cheese seller for ten years before starting up as a cheese maker in 1972: 'I discovered that there were only sixty-two Gloucester cattle left in the whole world. I couldn't believe it – that these wonderful animals should disappear simply because one or two other breeds of cattle had come to dominate livestock production. I thought the best way to save them was to breed them to use for the purpose they were originally bred for – making single Gloucester cheese. Gloucester cattle are lovely animals, with black heads and legs and a mahogany-coloured body, and of course they produce just the right sort of milk for our cheese.'

Charles is always happy to explain the difference between single and double Gloucester cheese: 'Single is more delicate and fragile and it is eaten young – at just eight weeks, in fact. Good double Gloucester wouldn't be ready that quickly.' Double Gloucester now completely dominates the Gloucester cheese market, but single may be making a bit of a comeback. Certainly things are not as bad as they were when Charles started work:

'At the time I got interested in cheese making, single Gloucester was officially extinct. However, there was a woman who, in the 1920s, had gone round the farms that

Opposite: A vat of soft, fragile, single Gloucester, a cheese that almost disappeared, but is now making a comeback among connoisseurs

still made single Gloucester – and there were lots back then in the 1920s. She watched how the farmers made the cheese because they did it all by instinct, as it were, growing their own culture and knowing when the milk was at just the right temperature for the various processes. When the farmers said it was at the right temperature she put a thermometer in the milk and wrote down the details. All this information was then published. I managed to track down a copy of her book, and better even than that, she heard that I'd started to make single Gloucester cheese and in the last months of her life, came to my farm to help. And that's how we got going – so without her the art of single Gloucester cheese-making would very likely have been lost for ever.'

So how is single Gloucester cheese made?

'Well, you warm your milk in a great vat – in our case it's in a 300-gallon vat – and then when it reaches just the right temperature, you put in your culture. The culture is basically bacteria which the old cheese makers would have grown every day. I buy it freeze-dried – my one concession to the modern world – because if you grow your own and make a mistake you've ruined a huge amount of milk.

'After the culture goes in, you leave the milk to stand for some time; then you add rennet. Next you cut it very gently with a special knife, a procedure that allows the whey to run off. You then scald the cheese to shrink it and to release more whey; traditionally the whey is fed to your Gloucester Old Spot pigs. You are then left with a thick curd that quickly knits together. This is cut and milled before salt is added. It's finally pressed into moulds, and then eaten eight weeks later. Easy, just so long as you know what you're doing – and nearly thirty years after we started I think we're still learning!'

Charles makes half a ton of cheese a week, which

Opposite: A tempting display of the range of traditional cheeses which Charles produces

Left: After rennet has been added, the cheese begins to separate into curds and whey. Traditionally the whey is run off and used as animal feed

sounds a lot until you remember that big cheese factories might produce fifty tons a day. 'We don't know how old single Gloucester cheese is,' says Charles, 'but it is mentioned in William Marshall's book, *Rural Economy of Gloucestershire*, which was published in 1796, and there's no doubt that it was extremely ancient by then.'

These days with refrigerated transport there are no problems with moving this most delicate of cheeses. As a result the reputation of single Gloucester is spreading, although Charles is usually the last to know where his cheese will eventually end up. He sells to wholesalers, and is always astonished when people tell him where they've seen his cheese – in Harrods, on the first-class decks of various airlines, and so on.

Double Gloucester is Charles's next plan, although he is a little reluctant to start competing on unfamiliar territory: 'The thing about double Gloucester is that it was never the cheese of the local people – I remember meeting two elderly ladies who'd spent the whole of their lives in the Vale of Berkeley and they'd never even heard of double Gloucester, although they'd eaten single regularly as children. It was the war that almost destroyed single Gloucester. Farmers were told to make only cheddar and one or two other cheeses that lasted and travelled well. The idea was to simplify. After the war the farmers carried on with cheddar, and single Gloucester was almost forgotten.'

Today Charles has two full-time cheese makers, and that threatened herd of Gloucester cattle has increased to a far more healthy seven hundred animals.

Left: Here the soft, crumbly cheese is removed from the vat before being taken off to the press
Right: Hard pressing marks the final stage in the cheese-making process – single Gloucester was originally eaten by the relatively poor

The Clay Pipe Maker

Right: A selection of hand-made pipes: William and Sue produce over one hundred different types

Opposite: William Young and Sue Lynch still make clay pipes as they were made in Elizabethan times

Tobacco arrived on these shores during the reign of Elizabeth I, and a flourishing pipe-making industry quickly grew up: such was the popularity of a weed which, in those far-off innocent days, was believed to be positively beneficial.

The first pipes – made at a time when tobacco was ruinously expensive – were carefully crafted from solid silver and these are now extremely rare. As the habit of smoking took hold and the price of tobacco went down, cheaper pipes made from clay began to appear. Clay pipes utterly dominated the scene for over two hundred years, but by the middle of the nineteenth century cigarettes and briar pipes were gradually being introduced – cigarette popularity increasing dramatically during the Crimean War – and slowly but surely the clay pipe began to be superseded.

From the heyday of clay, when some three thousand makers were involved in the trade, things have changed drastically, but astonishingly at least one company continues to make clay pipes. There is nothing gimmicky about this – it is simply that there is still a small demand for clay. The appeal is probably comparable to that of old pocket watches and snuff, both of which have their adherents, too. In the case of pipes the demand is supplied by William Young and his colleague Sue Lynch.

Based in a tiny workshop near Sheffield and using equipment and techniques unchanged for centuries,

William and Sue make some five hundred pipes a week. There are over one hundred different types and styles, but generally speaking, those with very long stems and faces on the bowls tend to go to historical organisations. Extraordinarily, many of the basic churchwarden-style pipes – those with stems 5in or 6in long – go to people who smoke them.

Sheila Kehoe, an elderly pipe-smoker who lives in a remote part of Ireland's County Wexford, explained the appeal: 'They produce a very clean smoke, and the clay lets the heat out well so the pipe itself stays cool. When I was young in the 1920s they were still common, but they don't last very long and so briar and other woods took over.' We may no longer approve of smoking – it is, after all, terribly unhealthy – but for good or ill some smokers insist, like Mrs Kehoe, that clay produces a more satisfying smoke than any other material.

The clay from which William and Sue make their pipes comes from Devon, and each pipe takes about a week to make; this includes drying and firing times.

The technique involved is a fairly simple one: the wet clay is rolled roughly into the shape of the pipe and then left to dry for a day at room temperature; a fine metal rod is then passed up the stem and left in position. The pipe is placed in a mould which is put in a press known as a gin. A lever on the gin is pulled to make the final, carefully moulded shape, complete with bowl. When the piercing rod is removed from the stem, any rough edges are smoothed off by hand and the pipe is left to dry for a day or two.

The final process is firing in a kiln. After firing, the stem ends are coated with wax or lacquer, an old but essential practice designed to prevent the smoker's lips sticking to the clay.

The Master Coppicer

NOTE Sadly, Bill passed away just before this book went to press.

Coppicing, the art of harvesting woodland by periodic cutting to ground level, has been carried out since ancient times. It is now rarely practised, but one man at least still sticks to the old ways: he is sixty-nine-year-old Bill Hogarth who learned his skills from his father and has spent his working life in the woods of the southern Lake District. He lives in Spark Bridge near Ulverston. Bill's contribution to his craft was recognised some years ago when he was awarded the MBE.

'I've been coppicing for fifty years and more,' he says. 'I started straight from school with my father during the last war, and his father, my grandfather, had been in forestry, too. Coppicing was a reserved occupation because the government needed thousands of ship's fenders, and they needed us to supply them. For the first ten years or so we used axes to cut the trees, but then the early chainsaws came in. They were good, but unreliable; today they are much better, of course, and what was almost impossibly back-breaking work is now much easier.

Opposite: Besom-broom in the making. A simple twisted wire grip keeps the head together and then the handle end is cut. Birch twigs have been used to make these brooms for centuries

'As well as the chainsaw, we also use a special peeler – like a chisel with a bevelled end – to strip bark. I work through about 15 acres of wood a year. In the depths of winter I make besom brooms from birch cut during the early winter – that is, birch is used for the brush part, and the handle is hazel.

'When you coppice a wood every last part of a tree is used: young hazel, for example, has many uses – for bean poles, fencing, walking sticks and, of course, for my brooms. Then there's oak: the bark from young oak trees is taken off, and the resin it yields is used for an essential part of the leather tanning process. Even the roots can be used – they're very popular with florists for flower-arranging bases!'

The bulk of Bill's work is carried on in winter, but in summer his time is taken up with the most evocative of all the coppicing skills: charcoal burning. Temperatures are critical for this process, and the slow-burning wood has to be watched twenty-four hours a day.

For many years Bill was the only person still coppicing in Britain, but in more recent times he has taught a number of people the craft, and his grandson, Neil, is now working with him. 'I hope that eventually he will take over from me,' says Bill.

'The wood we coppice goes into nearly two hundred products,' he says proudly, 'everything from didgeridoos to besom brooms. I get orders from all over the country, and I'm so busy I have to turn a lot of them down. It's lucky I survived, though! When I started, lots of coppicers were still about, but by the 1960s plastics were killing off the craft. For instance there was a huge bobbin mill near where I lived which used to take 5,000 tons of coppiced wood a year: then they turned to plastic, and you can imagine the effect that had on the coppicers. Swill basket makers began to close, too – there's just one left now that I supply. I managed to survive by supplying furniture makers, and in fact things have really picked up recently, what with the teaching and all.

'In winter we cut hazel for fencing mostly, and birch for steeplechase fences as well as brooms. May, June and July is peeling time, when we strip the bark for the tanners. Chemicals are used mostly for tanning, but the very best leather still needs oak bark – it's never been bettered. The poles left over go to the makers of rustic furniture. In September we start again on the birch and hazel.

'In three months' work with my grandson we might produce six tons of bark. When I started we got £8 10s a ton; now its over £3,500 a ton. Our bark goes to a tannery in Devon.

'Apart from the chainsaw and peeler, we use very few tools – a bill hook and a small axe is about all. I still use the peeler I started with more than fifty years ago!

'Some of our wood goes to pot-pourri makers, and the dead root of a coppiced tree – it might be five hundred years old – can be sold to people for fish tank decoration or flower arranging. Absolutely nothing is wasted, and it's all renewable. That's the beauty of coppicing.

'The tricky bit, and this is what I teach people, is that you have to watch your cycles. Hazel is cut every six to seven years, birch anywhere between eight and sixteen; the oak cycle is every twenty-five years, and alder, ash and sycamore anywhere between fifteen and twenty-five years.

'A wood I'm working in now has twenty-three species of tree, each with its own needs and uses. We used to work as far as sixty miles away, Dad and I, but now most of my work is within a few miles, largely because there is so much woodland and so few people to work it.'

Right: Bill Hogarth, a coppicer of the old school. 'When you coppice a wood, every last part of a tree is used.' He started coppicing straight from school, and worked for over fifty years

The bark peeler – a chisel with a bevelled end – produces a rich harvest for the leather tanners

The Fan Maker

Opposite: John Brooker, the last of the fan makers. At one time every woman carried a fan; then they started smoking cigarettes...

The majority of our work now is for film and television, but I see huge numbers of old and very rare fans, as well as making a new one now and then.'

John Brooker is Britain's only fan maker: 'Well, I'm the only one who puts "fan maker" on his tax return,' he says with a smile. Where once every woman from the lowest to the highest in the land would have had at least one fan, if not one for every occasion, they are now a mere curiosity.

John is fascinated by the history of the fan, but particularly the curious way in which it vanished from the social fabric of our lives: 'My own view is that the fan disappeared because women started to smoke – I know that sounds rather surprising, but I'm sure it's true. Right up until the 1920s every woman had a fan, from printed paper fans for the less well off, to incredibly expensive ivory and mother-of-pearl fans for the rich. But once the flapper girls came in, fans must have seemed a very old-fashioned encumbrance when you could be rather racy and have a cigarette in your hand instead.

'Mind you, it is surprising that despite the overnight demise of the fan as an almost universal item of dress, a few fan makers survived into the 1960s; the French firm Duvellerois was, I believe, the last to go. I think they had a shop in Regent Street until about 1963, probably serving a clientele born in the last century, and of course dying off with no new generation to replace it.'

John started making and restoring fans almost by chance. Through his wife's connection with a group of lacemakers, he decided initially that he'd make wooden frames in which collectors could display their fans. However, he quickly realised that no one was making the frames – or 'sets of sticks' as they are known – for the fans themselves.

'A fan is basically made in two pieces – the set of sticks, and the material that is stretched between them. Traditionally the real skill lay in making the sticks, because they had to fold neatly into a nest or be part of an elaborate telescopic or other arrangement. In eighteenth- and nineteenth-century fans, the ingenuity can be extraordinary; the lace or silk or gauze between the sticks was often beautifully painted, though this was usually by what we call jobbing painters who might copy an Old Master painting onto a fan, or the latest fashionable design. So-called "church fans" had hymns printed on their silk.

'We know that fans were used in ancient times – Tutankhamen's fan was discovered in his tomb and is now on display in a museum in Cairo, and I've seen parchment fans from the 1560s. I regularly get two- and three-hundred-year-old fans in for repair, and many are in

Opposite: Cutting a pattern in one of a fan's 'sticks'; one of John's biggest headaches is coping with delicate old fans that have suffered from amateur repairs

Below: A fan is made up of a set of sticks (right) and the material that is stretched between them (left)

excellent condition. Despite the fact that we don't use fans much today, people who've inherited them like to keep them and look after them.'

A set of sticks will take two weeks to make, unless they are highly decorated in which case they might take months; inevitably John is self-taught. He is meticulous in his work, though his theory is that much of the time he spends restoring a fan will repay him by teaching him about the early techniques of fan making.

'There are no specific tools for fan makers; when you set out to make a set of sticks you often simply start with a block of wood and cut it into thin strips which are then shaped. It can take a very long time indeed. I once tried a short-cut and used modern veneers which are obviously thin, but in fact they proved to be too thin.'

Folding fans came into Britain from the Far East in the middle of the sixteenth century, but by the eighteenth century they had become so popular they were available in a wide range of sizes, materials and styles: 'Oh, there are dozens,' says John. 'There's the cockade fan, the brise fan (which just had ribbon between the sticks), the telescopic fan and combinations of these styles and others.

'The most critical part of the fan maker's art is the pivot hole, which has to be exactly placed on the fan or it will not open and close properly. And repairing ancient fans is an even bigger headache.

'We see the most delicate, valuable fans with all sorts of terrible, amateur repairs, and it may take me weeks to work out how to undo what has been done before I can even start thinking about restoring the fan; but when you think that the record auction price for a fan is somewhere in the region of £15,000 you can see that repair can be worthwhile, even if it is expensive!

'My most satisfying moment was when a woman came to collect an ivory cockade fan I'd repaired for her. I'd replaced several sticks and she couldn't tell the new from the old, and she was so pleased to see the fan working again she was in tears.'

John recently made ten fans for the feature film *Elizabeth*, which he describes as being 'like giant bejewelled feather dusters'. His greatest regret is that he never went to see one of the last great fan makers, a man who died in Paris only very recently, when in his nineties. 'Imagine what secrets he would have had to share!' says John.

The Horse-collar Maker

Opposite: Straw provides the padding against which the horse pushes, while the checked wool lining absorbs the horse's sweat

Below: Stitching the pipe. The demand for horse collars is tiny today compared to a century ago, but the methods used remain the same

Now in his fifties, Terry Davis has spent all his working life making horse harness and collars – small ones for driving ponies and massive ones for the great Shire horses – but as he ruefully admits, the demand for horse harness and collars is not enormous.

'No, there isn't a huge demand, but in a way that doesn't matter because so few of us do it and it is so time-consuming that it's a full-time job supplying, say, thirty-five collars a year and the same number, perhaps slightly fewer, of sets of harness.

'I was trained as a saddler, but became more interested in collars and heavy working horses generally. After my apprenticeship I worked with an old man who'd made collars when working horses could be found on every farm in the country, so I had "insider" help, as it were. The point is that a horse collar is like a hand-made pair of shoes. You can't rush it, and each collar is made exactly to the measurements of a particular horse: no collar will fit two different horses, although a well-made collar will outlive the horse, its owner and probably me, too!

'Once you've got your measurements – and a Shire horse will vary between 25 and 27in – you make the pipe, or "forewell" as it's known, that runs round the front of the collar: this is a leather tube about 4in in circumference and it is stuffed with hay. Next you stuff the collar itself with straw. The collar is sewn together using conventional saddlery techniques, but what you have to remember about a collar is that although it's part of the harness, it is also separate in that it exists only to give the horse something to push against. It's really just a cushion, but vital to the whole operation.'

For driving ponies a collar would be completely leather lined, but for a big working horse, wool is the chosen material. Leather would make the horse sweat, whereas wool absorbs sweat.

'I send my collars all over the world – I've just had an order for a 32in collar from New Zealand. That must be

a vast horse – it's the biggest collar I've ever made. The working horse fraternity is quite small and everyone knows everyone else, so there's no need for me to advertise; it's all word of mouth. At the moment there's plenty of work to keep me, and perhaps two others nationwide, in business. I do the whole thing – I buy in the leather, cut it to size, stuff the straw in, add the attachments for the hames and so on. It's definitely a one-man business.'

The Hurdle Maker

Autumn and winter are the busy times for hurdle maker Michael Vincent, of Ilmington in Warwickshire, for he has to have a good supply of wooden sheep hurdles ready for the lambing season which has extended so greatly in modern times. Using traditional methods he can make an estimated six hurdles a day, with tools that have colourful names such as Tommy Hawk, barking irons and fromard, while the main aid to hurdle-making is a vice called a brake.

According to Michael the real secret is to make sure, when you are assembling the hurdles, that you 'leave enough room for the shepherd's foot' between the upright and the diagonal brace. Traditionally this was essential so the shepherd could push the hurdle into the ground using just his foot. Willow and ash are the timbers used for hurdle-making, but a shortage of ash means that Michael more often has to use willow. This is popular with shepherds because it is lighter, but willow hurdles are not as long-lasting as ash, and they take longer to make because the willow has to be barked.

Now in his sixties, Michael learned to make hurdles from an old man in the village of Coate in Oxfordshire: 'In the early 1960s there were hurdle makers all over the place and a big demand for them, but there are hardly any left now. It's all metal pens, although the woven type of hurdle is still made.

'I was originally a hedge layer, and that's what I did till about 1963 when the winter was so bad that we couldn't do any work. That was when I decided to start making hurdles. Originally it was to be just a wet day job, but gradually as hedge-laying went out of fashion it took up more of my time, although for a number of years I did both hedging and hurdle-making – hedge-laying in winter and hurdle-making in summer.

Opposite: Michael Vincent with his beautifully made hurdles; each one leaves 'room for the shepherd's foot' and is made by cleaving, not cutting, the timber

'Farmers used to buy a lot of hurdles – you might have had an order from one farmer for a hundred hurdles – but these days it tends to be smaller orders from what you might call hobby farmers, people with just a few sheep. Most of mine are used for lambing pens, because even the modern hi-tech farmer needs pens in his barns at lambing time. We also sell one or two to museums.

The cross pieces for a hurdle are fitted to the uprights in mortices cut with a Tommy Hawk

'I can't really tell you how long it takes to make a hurdle because I spend a few days making lots of parts and then a few days putting them together. Old Walter Long, the man who taught me the trade, wasn't far off the mark when he said that if you made six hurdles in a day you were doing well.

'Towards the end of my time – I'm largely retired now – I introduced a bit of mechanisation: a circular saw, a mortise machine and a small saw to make bark-trimming easier. But the traditional tools still have to be used where

awkward little jobs are concerned. For instance, the Tommy Hawk is an old-fashioned device for making mortises, the joint used to attach the rails to the "heads", as we call the end pieces. Then the barking iron is like a chisel with a sharp-edged 50-pence piece stuck on the end. It's used, as the name suggests, to strip the bark off timber.

'The fromard is an interesting tool: it's like a big metal L, where the short bit goes into the wood and the long bit is used as a lever to split or cleave the wood. All our rails are split rather than sawn. When you split wood like this you follow the grain round the knots and other imperfections and the wood retains its strength. Sawing obviously cuts through the fibres and so weakens the wood. This is one part of the work that can never be mechanised. You need a good eye to see which bit of wood will cleave, and then a steady hand to do the work properly.

'The brake is like a primitive vice that lifts and grips quickly so you can move the wood around as you work on it. We mostly use willow today as the ash supplies have dried up, though I'm currently using sweet chestnut which is excellent and very long-lasting – but whatever timber you use, it must cleave well.

'A good hurdle that is kept well covered when it's not in use will last many years, and although we hurdle makers are down to just a few, I don't think we'll ever completely die out. I suspect hurdles like ours have been made since at least the eighteenth century when the great boom in sheep began.'

Left: Here Michael holds the timber with a brake and pares it down using a traditional spoke shave

Right: Originally a hedge-layer, Michael has been making hurdles since the early 1960s

The Kipper Smoker

Opposite: Inside the Robsons' 130-year-old smokehouse at Craster. Here they produce some of the best kippers in the world

Below: Only a few boats remain in Craster, but the Robsons' fish are all locally caught

Some 40 miles north of Newcastle-upon-Tyne amid the wide, empty beaches of Northumberland, is the little fishing village of Craster. Its small harbour once had forty boats working out along the coast fishing for herring and lobsters and drift-netting for salmon. Along with fishing, Craster also once had a quarry which supplied kerbstones for London, using the harbour to ship them off. The lobster fishers and kerbstone makers have mostly gone now, but the Robsons are still there, and they intend to stay.

Most of the fish caught in the area traditionally went to the local merchants and to the smokehouse run by the Robsons, who nowadays obtain their fish from the three boats that still work out from Craster. The family has been in the business for four generations: their records go back to the early part of the nineteenth century, plenty of time to develop the sort of skills that produce what some say are the best kippers in the world. The Robsons are very particular about the fish they smoke in their 130-year-old smokehouse:

'We use only the plumpest herring with good oil content to produce our kippers,' says Neil Robson; Neil runs the company with his father Alan. 'It takes about eighteen hours before the kippers are ready, and we use only oak shavings – nothing else produces the right flavour. The fish are split, and then hung in rows high enough above the smoke so they don't cook.'

The Robsons believe that the decline in sales of smoked fish from small producers is the fault of EU quotas, and numerous rules about hygiene: 'There is no doubt that things have become more difficult for the smaller producer, because it's easier for a small company like ours to fall foul of rules that are only really appropriate to companies producing food on a massive scale – which, of course, we don't. But we still have a niche in the market with quality products sent direct by post to the public, and deliveries to smaller retailers. Our kippers are expensive compared to mass-produced ones, but we don't use artificial colourings or any chemicals, and we do use the best fish, the best oak, time and skill. Kipper connoisseurs seem to think that paying a little extra is well worth it!'

23½
E&P

The Legal Wig Maker

Throughout the seventeenth and much of the eighteenth century no self-respecting gentleman would ever have considered stepping outside his house without his wig. Wigs were simply part of elegant dress, as essential to the man of fashion as a jacket is to a modern suit. By the nineteenth century, however, wigs had all but disappeared, except among barristers and judges, and the legal profession has held fast to the tradition of wig-wearing ever since.

The traditions and techniques involved in making a legal wig are much as they would have been for any seventeenth-century wig, and the one wig-making firm that survives to this day, Ede and Ravenscroft, can trace its origins to 1689. At that time they were robemakers; then in 1726 Thomas Ravenscroft, a wig maker from Shropshire, moved to London and began making wigs for the church and the law. His grandson Humphrey invented the barrister's wigs we know today: these are known as forensic wigs. Though similar in essentials to earlier wigs, the forensic wig removed the need to constantly curl and dust, both of which procedures had been an essential part of earlier wig-making. Humphrey's forensic wig is still in use today.

Originally Ravenscroft traded in Serle Street, London, but in the 1890s a daughter of the firm, Rosa, married Joseph Ede, the son of a well-known robe maker, and thus the company that we know today came into being. It was at this time that the company moved to 93 Chancery Lane, in the heart of London's legal district, where they remain to this day.

Legal wigs are made out of horsehair, and vary according to who will be wearing them. The acknowledged world authority on wig-making today is Kathleen Clifford. Kathleen, now in her sixties, has been making legal wigs at Ede and Ravenscroft for nearly fifty years: 'I came here as a girl of fifteen, and I've been here ever since,' she says with a smile. 'We are the only legal wig makers left in the world. You can't learn how to do it at college or anywhere else: it's here, or nowhere.'

Kathleen and the three other wig makers currently employed by Ede and Ravenscroft make something like a thousand barristers' wigs a year, about 120 bench wigs (a judge's working wig) and about a hundred full-bottomed wigs (worn at ceremonial occasions and only by judges). That may seem a large number, until one remembers that Ede and Ravenscroft are the only people still in the wig-making business.. The wigs are all made on wooden blocks, each shaped, obviously, like a head. Many of these are more than a century old, but they are still perfectly serviceable. Outworkers – usually women from Ede and Ravenscroft who choose to work from home – weave the horsehair which is then sent in to Kathleen and her fellow workers. The woven horsehair is stitched on a silk netting base – and this is where the skill comes in:

Opposite: Wigs in the storeroom at Ede and Ravenscroft, the last makers of legal wigs in the world

'It takes about a year to learn how to sew the hair on – obviously it has to be neat, with tight curls and with no gaps, and this is not an easy thing to achieve when you're working with horsehair; but it's all still done as it was centuries ago. All the horsehair is curled using old-fashioned curling tongs which have to be left on a fire first. For pressing we use irons that also have to be heated on a fire. It's all wonderfully Dickensian! But our wigs go wherever wigs are still worn, which means mostly Commonwealth countries. We are the very last of the makers.

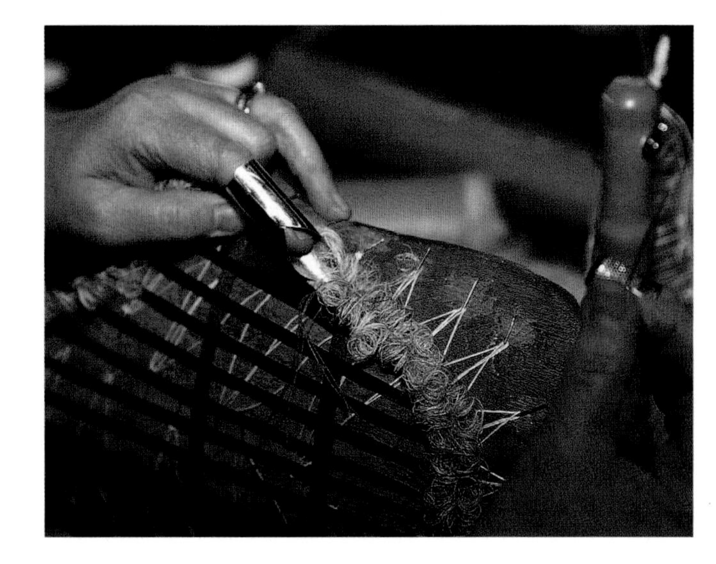

Right: The skills involved in wig-making are pretty much as they were during the eighteenth century, a time when no gentleman would be seen in public without his wig

Far right: Kathleen starts work on a silk netting base, the material on to which the intricately woven horsehair is stitched

'It takes three or four weeks to make a barrister's wig, and about twenty-four hours to weave enough horsehair for one wig. We take five measurements when we're making a new wig: the circumference of the head; the distance from forehead to nape; the circumference over the head; and the distance from ear to ear at the back.

'Most people assume that because the wigs are made from horsehair they'll be heavy and uncomfortable to wear, but in fact they are very light – and of course, they last a lifetime.'

The Mud-horse Fisherman

It is difficult to imagine anyone making a living from the treacherous mudflats of the Bristol Channel, yet this vast, hostile area is home to one of the rarest and most unusual trades left in Britain: the mud-horse fisherman. It is difficult to be sure, but legend has it that before the Romans came to Britain the Celtic tribesmen who lived in this area used a strange wooden sled to propel themselves across areas of mud that would otherwise have swallowed them up in an instant. The sled was probably developed originally when it was realised that nets set far out on the mud at low tide were highly productive when the waters rose, bringing fish up the estuary and, with luck, into the nets. These fish were then collected on the next low tide.

Skill and experience handed down from generation to generation, combined with the weight-bearing properties of the mud-horse – as the strange sled came to be known – enabled a rich harvest of fish to be collected regularly from an otherwise inaccessible environment. Only one mud-horse fisherman still plies his ancient trade: Brendan Sellick's family have been mud-horse fishermen for well over a century, and the chances are that the first family member to learn the trade did so with a man who would have learned it in a direct line stretching back centuries.

When Brendan started more than fifty years ago there were still a dozen or more mud-horse fishermen working the mudflats on the south coast of the channel. All except Brendan are now gone.

The mud-horse Brendan uses is as old as the century, and it is a remarkable piece of equipment by any standards. Rumour has it that the wooden boards from which it is made came from the local undertaker's. Whatever the truth of that rumour, it is simply a sled with a running board and handles high enough to enable the rider almost to stand upright. The sled can actually be pushed across the mud far more rapidly than one might imagine… Brendan sets off for his nets at a cracking pace, one foot on the sled, the other pushing back into the mud.

He takes visitors out the one mile to the nets over firm mud so that they can see how the mud-horse is worked while following behind on foot. But he travels alone out over mud so treacherous that without the mud-horse he would be sucked under forever in seconds. It saddens him to think that he is possibly the last of the line; as he observes:

'In the past, fishermen on many estuaries used a mud-horse to get about – they certainly used them up in Cumbria and along the South Wales coast; but I am the last one left. My only hope is that my son will take over –

Opposite: Brendan Sellick on his mud-horse. Brendan ekes out a precarious living on the treacherous mud flats of the Bristol Channel

Opposite: Tending the nets: when Brendan finally retires mud-horse fishing may come to an end after nearly two thousand years

at the moment he is keen, so the trade may last a while longer.'

Anyone who has gone out with Brendan is immediately astonished at the speed with which he travels. 'I just couldn't keep up with him,' said one local journalist aged twenty-five, 'and I couldn't believe it when he said he was nearly seventy!' Brendan may travel a mile and more out to the nets where he may catch anything from shrimps to cod in nets strung between special poles.

'Tidal fishing is practised here in this way because of the huge tidal range,' he explains. 'I've been doing this since I was fourteen, and I can't see myself ever stopping. Despite the cold and the exhaustion and the fact that we don't make huge catches – so I'm not going to get rich – there is something about the life that gets into your blood. I'll probably carry on until I drop.

'I rarely return empty-handed, but there are fewer fish than there used to be. At least no one can say that this is not an environmentally friendly way to fish. If everyone fished the way I do, stocks would never have got depleted – and they are seriously down. I've noticed a drastic decline over the years.'

Brendan sells most of his catch to local hotels and restaurants, and then returns to his seaside cottage with a few fish – sometimes a salmon, or a few eels – for his own supper.

The Peat Cutter

The Slievenanee mountains in Northern Ireland's Glens of Antrim are rich in peat, and for centuries each family in the area had rights to cut in certain specific places. These rights were jealously guarded, but with the coming of modern fuels the peat cutter – the man who still cuts peat by hand – has become a very rare creature indeed. However, one man at least continues the ancient tradition: William McBurney cuts on what is known locally as the moss, near his home in Newtowncrommelin. William has made few concessions to the modern world: the floors of his house are carpetless, he would never dream of installing central heating and he still speaks in marvelling tones at the wonder he experienced the first day he found that he could bring a transistor radio out onto the bog and listen to its voices.

William lives alone now and therefore cuts less peat than he did as a young man with a house full of brothers and sisters and, of course, his parents. His grandfather bought the house – a former schoolhouse built by Huguenot settlers – nearly a century ago, and little has changed in it in that time; most significantly, it still has the old range where the sweet-smelling peat burns every day. According to William, peat-cutting was once very much a social affair with the whole family taking part and meeting other local families out on the moss. When he was a boy, William would ride with his uncle on a wagon down to the bank, carrying his father's lunch – a soda farl griddled by his mother, and a tin flask of milk.

'There used to be people living in nearly every house along our road,' says Mr McBurney, who is now in his sixties; 'now there's just one or two. It can be lonely. You meet people at church, at the local markets and at sheep sales, but out on the moss at peat-cutting time I now see almost no one at all.' The old peat banks cut from the side of the mountain can still be seen here and there throughout the district, but most are long abandoned. They show like black bites out of the hillside.

Opposite: Peat left to dry in the summer sun

Below: William McBurney with a great store of the wonderful fuel

Opposite: Peat-cutting was once a social affair with whole families taking part. Now only a few diehards continue to cut in the old way

William's rights to the peat go back a long way, to the days of Nicholas de la Chevis Crommelin who owned the land in the seventeenth century. The original rights, which are unchanged to this day, state that anyone living in the three townlands of Skerry East, Skerry West or Scotch Omerbane has a right to cut peat to burn in their own holding. There was once a strictly enforced ban on cutting peat to sell, but so few people cut turf by hand nowadays that it would be unlikely to arouse anyone's indignation.

The curiously shaped peat spade used by William McBurney is unique to the Ballymeena area

William opened the peat bank he still cuts, with his father, in 1956. 'There's over forty years' work here,' he said, lifting the locally designed, Ballymena-pattern peat spade that was once used throughout the district. In the old days someone would have ferried the peat in a barrow and put it into heaps, ready for being spread, and once it had begun to firm up in the sun, would have carefully built it into space-saving castles. Now he performs both tasks himself, swiftly building tent-shaped piles that enable the peat to dry quickly.

Modern methods have begun to have an impact on the moss, which is home to rare plants and birds, including butterwort and skylarks. A bank has been cut by mechanised diggers a little way down the slopes of the mountain, and that, explains William, could take away a bog in two or three years. 'By hand it could go on for centuries,' he says. 'It's the machines that do so much destruction – they clean everything up so quickly. That's why there's so many complaints about it. The bogs will disappear altogether.'

The bog is also home to a way of life that dates back to the times when crofters built their now-ruined cottages. Some kept sheep or one or two cattle, some mined for iron ore in the days when Crommelin founded an ultimately unsuccessful smelting works – but all would have cut peat. 'There's nothing but old, vacated, tumble-down houses,' says William. 'A lot of people say to me, "I don't know how you live there!", but I've lived here all my life. I can't imagine anything else, and anyway, it is a very beautiful place.'

Even older stories can be read in the peat, in the preserved remains of a forgotten landscape: 'Quite often you find ancient fossilised pieces of wood,' says William. 'They're so hard, it's easy to break your spade on them – they're like stone.' He sent one of these old scraps of wood to the Ulster Museum, and was told it was a 3,000-year-old piece of Scotch pine. These fragments are all that remain of trees that grew when the surface of the bog was dry and firm.

William doesn't believe that the life of the hand peat cutter will completely die out, but he does admit: 'It's definitely on the wane: most people go for machine turf now, and a lot have oil-fired heating.'

The Pole-lathe Man

From archaeological evidence we know that some two thousand years ago pole lathes were being used in England to cut and turn bracelets from shale. Moreover in Egypt and other Third World countries pole lathes exist to this day that are, in their essentials, identical to those used in Britain for a thousand years. In Britain the real home of the pole lathe is the area around High Wycombe in Buckinghamshire where English chairs were traditionally made. Here, extensive areas of beechwood – the staple of the chair-making industry – have always been available to the chair and pole lathers, whose numbers increased dramatically as the nineteenth century progressed.

The vast increase in the urban population living in small houses created a huge demand for cheap wooden chairs, and High Wycombe and other similar areas expanded to meet the demand. Today there are few traditional chair makers left in the area, but one or two 'bodgers' still ply their trade in the woods here. The chair bodger was the man who made the turned parts of the chair, namely the spindles for the back, the legs and the cross-pieces. Bodgers in the High Wycombe area rarely made the seat or completed the whole chair, but a related tradition in Herefordshire and other rural districts saw craftsmen turning their hand to seats, poles, backs, spindles and assembly – the whole thing, in fact.

In Herefordshire today, thanks to the efforts of a number of individuals including Mike Abbot, the pole-lathe chair maker is still with us. True to the local tradition, Mike makes his chairs himself from start to finish. He is also a mine of information about the history of chair-making and bodging:

'Pole-latheing nearly died out – technically it died out completely for a few years – because industrialisation gradually made the job redundant. It was the Arts and Crafts movement at the end of the nineteenth century which delayed its demise and kept it going into the 1950s. As most people know, this movement was a sort of reaction against the machine age and supported the continued production of hand-crafted furniture, textiles, pottery and glass. Pole-lathe work appealed to the desire for individual skill.'

It is difficult to describe exactly how a pole lathe works, although to see one in action is to realise it is simplicity itself. A springy sapling is fixed in the ground and attached at the thin end, which should be above the worker's head, via a cord to a lathe. When the lathe operator pushes his foot down on a pedal the cord is pulled, the pole bends down and the lathe turns. When pressure is released off the pedal, the springiness of the pole turns the lathe in the other direction by pulling upwards. All the while the craftsman's various chisels are cutting and shaping the turning wood.

'Apart from the pole, the lathe is just a simple wooden frame,' says Mike. 'I make my own – it would take anyone that was competent only about a day to do it – and I use a pole of alder or, better still, ash.

'When I started there were just half a dozen of us hanging on, compared with thousands doing it a century ago. I

Opposite: Mike Abbot in action: pole-lathed timber is slightly oval in section and each piece is unique

started out doing tree surgery and gradually got interested in pole-latheing and chair-making. Then I saw an old film of the last pole lathers, part of a living tradition, as it were; they'd been filmed in the 1950s specifically because it was felt, with justice, that their skills would soon be consigned to history. I learned a lot from watching that film and from reading old books – enough, anyway, to keep the craft alive.

'Like the chair makers of old I still use planks for the seats of my chairs, rather than strips of wood glued together as factory-made chairs tend to have. Apart from the pole lathe I use an adze – a medieval tool that was traditionally used to scoop out the shapings in the chair seat. The rule was always: the bigger and deeper the scoops, the better the seat.'

These days Mike teaches chair-making and pole-latheing, saving it from extinction – for the time being, anyway. Traditionally pole-latheing was winter work – chair bodgers did other things in summer – but for Mike the business is an all-year-round one. But has he ever felt the need to cut corners, to turn his wood on an electric lathe?

'Well no, because the great appeal of the pole lathe was that it was portable – you went to wherever the wood was, with your lathe carried over your shoulder, then just set up under a suitable sapling and set to work.

'I make traditional chairs, mostly of the Windsor chair variety, but I also design my own, especially various kinds of ladderback. But whatever the chair, the poles for it – the pieces for the back, the legs and various cross-supports – are always made on the pole lathe. An expert might know they were not machine-made by the fact that pole-lathed timbers are always slightly oval in section.'

To ensure a ready supply of mixed timber, Mike and a group of friends have bought their own piece of woodland near Ledbury in Herefordshire. From this 100-acre wood they are able to harvest regularly on a ten-year cycle.

Left: Pole-latheing in the Herefordshire woods. The great advantage of this ancient wood-turning system is that it can be set up almost anywhere

The Pub Sign Painter

Volumes have been written on the history of the pub sign. The Romans used to hang a bush outside their taverns, from which we derive the phrase 'good wine needs no bush'; Pliny mentions a tavern called The Cock, as well as a wineshop known as The Shepherd and Staff – a name with a decidedly modern air to it.

In medieval times when few could read, all shops had signboards: vases of perfume outside a perfume maker's, an adze and other tools outside a carpenter's, and so on. Taverns probably inherited the tradition of signs from the nobility whose houses were often used as hostelries for travellers while the family was away – the coat of arms that hung outside the house gradually transferred to public houses and taverns, so for example what on the mansion had been an azure lion became in popular parlance the 'blue lion'. Coats of arms became increasingly popular, and red lions and green dragons sprang up everywhere. The name of a pub in Lewes, Sussex, known as The Three Pelicans, had its origins in the arms of the Pelham family.

Ale house signs eventually became so big that ordinances were issued limiting their size. By the twentieth century the range of pub and inn signs had proliferated to include sporting motifs, commemorative signs, pagan and heraldic devices and, more recently, odd combinations that appeal solely because they are odd, such as The Slug and Lettuce.

It is regrettable that in this world of pub sign painters and makers we know very little about the men and women who made and painted them. Pub signs were not signed, and only in very rare circumstances have individual signs survived more than a few decades. But this ancient trade is still carried on by a few individuals, and one of the most experienced still with us is Mike Hawkes who has been painting pub signs for nearly forty years. Mike trained as an artist, worked in an animation studio, and then went to the now defunct West Country Breweries who then employed three full-time sign painters.

'It's been a wonderful job,' he says with a gleam of real pleasure in his eyes. 'You get the chance to be part of an ancient tradition, and your work is seen every day by thousands of people. You also get to paint virtually every subject under the sun, other than the really modern stuff, which I'm not interested in anyway.

'We don't do slugs and lettuces, but we do countless coats of arms, red lions and portraits of kings and queens. And since I retired as an employed pub sign painter and went freelance in the early 1990s, I have only done signs on a purely commission basis, so I get to pick and choose what I do – most of those who commission me now want to keep the signs as paintings rather than hang them outside. I'm currently working on a board for an American who wants it to be packed with things typically English: a red phone box, fox hunting, sheep, old houses. My decades

Opposite: Mike Hawkes with a half-finished sign. Though the craft of pub-sign painting is centuries old, individual signs rarely last more than a couple of decades

of experience in painting pub signs is the perfect background for such a tricky job!

'With pub work the details tended to be left to me, so I wasn't always doing exact copies of the worn-out sign that had been hanging there – and that's part of the tradition, too. Artists were always given plenty of licence to adapt and change, or to add their own little flourishes.

'Most people think pub signs last for ages, but they don't, even when they are carefully prepared, as mine are, and painted with endless coats of underpainting and careful sealing – they'll still only last fifteen years at most. The south side is usually the first to fade because of the amount of sunlight the picture is exposed to. But the wind and the rain eat away at it, too, so however well it is painted and varnished it will inevitably have to be painted again. Several times I've gone back to repaint a sign I originally painted no more than a decade before.

'One sign takes at least three weeks to paint. Most of this time goes in preparing the surface for painting – and this is where amateurs go wrong, because without good preparation the sign may last only a couple of years or even months. And, of course, we have to paint both sides – don't forget that!

'As I say, signs rarely last more than a couple of decades at most – although in my collection I've got a few dating back forty years. Some that are centuries old exist in museums, but the fact that they have survived so long is almost certainly because they have been thrown into an attic and forgotten. Anything left hanging outside a pub deteriorates.'

Mike is now in his seventies, but he has no plans to stop painting: 'While the work keeps coming in, I'll keep doing it,' he says. 'And who knows – my work might last a bit longer, as I always sign everything I do. Not like the pub sign painters of the past who are sadly unknown to us.'

Opposite: Mike with a pub sign that combines many traditional English images. The secret of good pub-sign painting, he explains, lies in the preparation

FOX & HOUNDS
MIKE HAWKES 1996

The Putcher Fisherman

Opposite and overleaf: A scene of extraordinary contrasts: in the shadow of the ultra-modern Severn Bridge, the old putcher frames still provide a living for a few hardy fisherman

Below: John Walters with the day's catch

John Walters is one of the last putcher fishermen. The origins of the word 'putcher' are unknown, according to the *Oxford English Dictionary*, but the word is related to the dialect word 'putchen', meaning an eel basket and was once common across Shropshire, Worcestershire, Warwickshire and Gloucestershire. But whatever the origins of the name, the putcher fisherman has for centuries been unique to the River Severn.

'They have fished this way since Saxon times, but only on the Severn so far as we know,' says John who has been putcher fishing since the early 1960s, 'but it's in danger of dying out.' The problem is that the local authority is about to reduce the length of John's season to ten weeks, which may mean the business will become uneconomic.

'I think it's madness,' says John, 'because the salmon stocks seem healthier now to me than they were in the early 1960s.' John started as a putcher fisherman after a long wait: 'Well, putchers are fixed in a large framework which is built out from the bank. There were always only a limited number of these fixed frames and you had to wait until one came up for rent – when I started it was a question of dead men's shoes. We rent ours from the Duke of Beaufort who has probably owned them and made money from them for centuries.'

John eventually looked after three sets of putchers, but in recent years he has only worked the one at Beachey, near Chepstow. 'As I now work on my own, one set is the most I can manage because it's a lot of work looking after a putcher frame. Boats smash bits of it regularly so it's always having to be repaired, and you have to visit it several times a day to collect the fish. If you don't get there quickly every low tide, someone else will.'

The putcher frame is made from larch poles up to 40ft long, nailed together in such a way that nearly eight hundred putchers – the nets – can be fixed in rows; at high tide they then fill all the water between the surface and the

bottom. The framework on which the nets are fixed might run out from the bank some 350ft and be as much as 25ft high. The idea is that salmon try to swim through the wire-mesh putcher and get stuck. Each putcher is between 24in and 30in in diameter at the mouth, and about 4ft long. Many are still woven from willow, but metal lasts longer.

'We occasionally catch other fish,' confesses John, 'but we really are only after salmon, which fetch about £6 per pound in April and £2 to £3 per pound in June. However, if we have to start on 1 June instead of 16 April – we finish on 15 August – and are not given a drop in the rent we pay, I think we'll have to pack up, which would be sad. But there it is.'

At low tide John walks along the top of the putchers, picking out any trapped fish. At the beginning of the close season he takes down the eight hundred-odd putcher nets and the rails that make up the frame; this allows him time to repair and replace anything worn or broken. But in some years there might be a disaster early on in the season, as he explains:

'Several times we've had large parts of the frame smashed by a boat or a tree right at the start of the season. This means you can't fish at all, and you don't get your rent back. Moreover, quite apart from the financial difficulties, putcher fishing is very hard work physically. All the maintenance and care is by hand, and a putcher frame is big and very demanding.'

Here and there along the Severn are the remains of once-flourishing putcher frames, because at one time this type of fishing was far more common. John originally revived a putcher frame that hadn't been used since the early 1800s; but sadly it looks as if new rules and regulations may soon put an end entirely to the last of the putcher fishers.

The Rake Maker

Back in the days when farm implements were all made locally and from local materials, enormous ingenuity was often required to make a serviceable tool from limited resources. But in turn this often produced a situation where, over centuries, tools developed that were so well suited to the task in hand that they have never been superseded even by tools made with the latest hi-tech materials.

A case in point is the garden rake. It might seem that a metal rake would work better and last longer, but in fact wooden rakes – probably made in exactly the same way for centuries – are actually more comfortable and efficient to use than modern metal ones. Traditional wooden hay rakes are now very rare indeed, but we found one man who is still making them to a design that would be recognisable to an eighteenth-century farmworker: in the tiny hamlet of Dufton, near Appleby, in Cumbria, the Rudd family have been making wooden hay rakes for over a century. Working in the late seventeenth-century workshop that was used by his father and grandfather, John Rudd – now in his late sixties – and his son Graeme currently make some 14,000 rakes a year, still using the tools and techniques that were used when the business was started by John's grandfather in 1892.

Hay rakes are made from three timbers: ash, ramin (a hardwood imported from the Far East) and silver birch. The ash and ramin are used to make the head and shaft

Opposite: John Rudd in the rake workshop set up by his grandfather over a century ago. Though wooden rakes nearly disappeared in the 1960s they are now hugely popular

John prepares to nail the hoop to the head and shaft. It's the hoop that gives the rake its real strength

Opposite: John with the finished article: he makes several hundred rakes each week

of the rake respectively; the sixteen teeth are always made from silver birch. Traditionally pitch pine was used for the handles, then earlier this century Colombian pine started to be used; but good supplies of straight-grained wood began to dry up, and John moved to ramin. Rake handles are a standard 6ft long, and the whole thing is hand-made except for a machine that pushes the teeth into position – this is purpose-made, and therefore unique.

John says he has been making rakes since he could walk: 'I'd have been about ten, anyway!' he says with a smile. 'It was what the family did, and I learned it the way most kids learn to play football; but even then I knew we were doing something special. I doubt if anyone else in the country still makes them the way we do.'

So how are these splendid rakes made? Surrounded by wood shavings, John was happy to explain: 'First you shape the 6ft-long shaft; this needs to be straight and smooth, obviously. Next a small wooden ash hoop, about ½in in diameter, softened by boiling for ten minutes and then shaped into a semi-circle, is used to hold the head onto the shaft – as it dries out and hardens, the hoop grips like a vice, but it is also nailed. A traditional rake involves the use of just four nails – three in the hoop, and one into the shaft.

'Birch has always been used to make the teeth because it is plentiful locally and hard-wearing, and the latter is, of course, essential since the teeth do all the work. My grandfather told me that birch used to be made into clog soles, so it must have been pretty solid stuff. With care, a rake will last twenty years. I don't know how long it takes to make one because we make lots of hoops and heads and shafts, and then fix them together. But we might make a few hundred in a week.

'Originally most of these rakes would have been used for hay, but these days people use them on farms, in their gardens, and even as shop-window ornaments; and a lot of people buy them because they like the look of them because they are hand-made. But they are also popular for raking over bunkers on golf courses. I see our rakes used on the greens in top golf tournaments and sometimes in period television dramas, which always gives me a good feeling.'

John sells most of his rakes to a wholesaler in Sheffield, and from there they go all over the world. 'It's wonderful to make something by hand that's so popular,' he says. 'Of course, it's also useful and it's part of a long tradition.'

The Rhubarb Forcer Maker

It was a chance inquiry made by the Duke of Edinburgh at the Chelsea Flower Show some years ago that led potter John Huggins to start making giant terracotta rhubarb forcers. 'He asked me if I made a really good, big rhubarb forcer,' says John,'and I had to say no, but it made me think.'

Now he makes hundreds a year, all by hand in his little workshop in Ruardean in the Forest of Dean; in fact he is the only man left making hand-made forcers. 'I reckon I must also be the biggest producer in the country, if not the world,' he says proudly. 'I sell them all over; gardeners like the look of terracotta with its strong red colour.'

Each forcer starts life as a massive, wet lump of clay weighing about 45lb which is thrown on John's potter's wheel: 'It takes time to work the clay up into the shape I need, and it's difficult to say exactly how it's done – with plenty of experience you get a feel for it. But 45lb of clay is quite a lump, and if you don't know what you're doing and you don't get on top of it quickly once it's on the wheel, it will quickly get on top of you – or all over the workshop, anyway!

'After firing, the forcer is frost-resistant, but the fact that it is able to resist cracking in this way also has a great deal to do with its being hand-made: when you make it by hand you create a sort of spiral effect within the clay that gives it resistance. Pots made by machine tend to have layers of clay because the machine just slams into the clay to make the

Opposite: John Huggins in his workshop: each forcer starts life as a 45lb lump of clay and is made entirely by hand

Left: An imprinting wheel adds a guarantee of quality to the finished article, telling each buyer that this is a unique hand-made forcer

walls, and so cracking in very cold weather is far more likely.

'We have three people now working full time making the forcers and other pots.'

John used to be a lexicographer – a professional dictionary writer – but he admits with a grin: 'I gave it up because it was so boring! This was about twenty years ago, and I'd always loved clay pots, so it seemed a quite natural step to go into them. For the forcer, we started by copying an old clay rhubarb forcer design – you used to see these big rhubarb forcers all over the fields where rhubarb was grown commercially before more intensive methods came into use.'

When rhubarb was first introduced to Britain in the seventeenth century it was rare, expensive and highly valued for its medicinal properties. By the nineteenth century it was very popular simply as a pudding ingredient, and as the demand for rhubarb increased it was discovered – necessity being the mother of invention – that a terracotta forcer improved yields considerably.

It takes only about twenty minutes for John to turn his 45lb lump of clay into a forcer, but it then has to be left for three to four weeks to dry before being fired slowly for three days. As for the clay: 'We get our clay from near Bristol and bring it here raw, which means we know exactly what we're getting – again, that's important when you're trying to make something of quality. Poor-quality clay will produce poor-quality pots. Then you have to fire your forcer slowly, and afterwards cool it slowly, or it will break,' says John.

The time involved in the making of each forcer means that John makes only a few hundred a year, but they are very popular: 'We had a big order from the Japanese. However, many people here buy them just because they like the sort of monumental sculpture quality – after all, they are nearly 2ft high. I'm sure many people don't even grow rhubarb in them!'

Opposite: Using a simple straight-edged tool, John smoothes the outside of a forcer. After this it will be left to dry for several weeks before being fired

Left: Huggins rhubarb forcers are keenly sought by everyone from practical gardeners to interior designers

The Ships' Figurehead Carver

Steve Conway carves ships' figureheads at his studio near Barnstaple on the edge of Exmoor. He specialises in the conservation of painted wood, but over the years he has become increasingly fascinated by the history and traditions of ships' figurehead carving, conservation and restoration. The nineteenth century was the great age of the figurehead: thousands of new ships were commissioned and built each year as the British Empire, and with it British trade, expanded round the world – and no sailing ship was complete without a carved figure at the prow. But where does this curious tradition stem from? Steve explains:

'The tradition of having a figure at the front of the ship dates back almost beyond recorded history. Certainly the Egyptians carved figures on the front of their ships, particularly funerary ships – we know this because their tomb drawings make it abundantly clear – but figures have been carved on the front of ships in every age since then, and in most sea-faring cultures.

'The idea behind the figurehead we think is that it first and foremost protected the ship and its crew from the perils of the sea – a sort of good spirit, if you like: the figurehead was also out there at the front seeing, as it were, the way forward. In Greece this tradition, or a form of it, is carried on to this day in the occulus which is painted on even the smallest Mediterranean fishing boat. The occulus, basically an eye, protects the boat's crew from evil. By the seventeenth and eighteenth centuries most European maritime nations were carving lions as figureheads – the lion was dominant throughout European navies and for obvious reasons: it was the king of the jungle, the fiercest and most fearless of all animals. The idea I suppose was that some measure of its courage and strength would somehow rub off onto the ship to which it was attached.

'In the sixteenth century we know that a much wider range of creatures were carved as figureheads – the *Golden Hind* with its carved deer is a good example. But they were also carved with gryphons, unicorns, elaborate classical figures and so on. Sadly very few figureheads survive from this early period. In the twentieth century the tradition has vanished almost entirely, along with the tradition of carving nearly every other bit of the ship – now, the passion for ship's decoration which was characteristic of the era of wooden ship building has been replaced by a passion for pure efficiency which, of course, is the central aim of the modern ship.

'Concern about costs eventually reduced the amount of carving on the old wooden ships, but a wooden ship without a figurehead must have been unimaginable. Steel and iron ships didn't need superstition when they had extraordinary engineering to rely upon.'

Steve originally helped restore old figureheads, but his interest grew, and now the bulk of his work is making replicas of original figureheads which are so fragile that they

Opposite: Steve Conway poses with a magnificent, if half-finished, figurehead: on a first-rate ship of the line – a ship with a hundred guns – the figurehead would have been at least 10ft high

Opposite: Carving a figurehead involves crude early cutting – to get the rough outline – followed by delicate work to produce the fine detail of hair, drapery and features

cannot be kept outside or on display in any unstable environment. The irony, as he points out with a wry smile, is that he has almost never worked on a ship's figurehead on a ship:

'The vast bulk of them are in collections, their ships long ago broken up. Timber ships were never expected to last long, and there was no sentimentality about them until recently. If you recall Turner's picture *The Fighting Temeraire* you will remember that it shows a famous ship from the Battle of Trafalgar being towed away to be broken up. Turner may have seen it as a tragedy for a warhorse that should have been kept as a national treasure, but no one else did.'

The tradition was to sell the ship's timbers for house building, furniture-making and so on; the figureheads, however, would often be bought and kept, and today they are much treasured in collections round the world. But inevitably their paint peels and their timber decays, and this is where Steve comes in:

'When I'm asked to replace an old figurehead that's on a ship I never make a cast. I carve an exact replica by hand, just as the original would have been carved, and probably even using the same tools.' These days commissions for new figureheads are in fact far more likely to come from an interior designer than from a shipbuilder – but at least the craft remains a living reminder of a glorious past. The early figureheads were carved in oak or elm, but by Victorian times Western red cedar was extremely popular as it is resistant to attack by fungus and woodworm. The *Cutty Sark*'s figurehead was re-carved in red cedar in the 1950s.

'The real skill comes in carving the hair, the drapery and the other details,' says Steve. 'On original figureheads these details tend to get obscured by numerous layers of paint, so when I'm restoring something I go down through the layers very carefully to find out how it was originally painted.

'From scratch is takes about three months to make a life-size figurehead. On what was known as a first-rate ship of the line – that would have carried a hundred guns – the figurehead would have been at least 10ft high, a massive feat of carving skill. I've worked on figureheads this big at the National Maritime Museum where there is a huge collection.'

The rarest figureheads are those originally carved for merchant ships – in the nineteenth century many of these would have had the figure of the ship's owner, or his wife or daughters on the front. The most unusual figurehead Steve has worked on was carved for a ship built in Bombay, again in the nineteenth century which depicted a man sitting on the back of a giant eagle with his bearer or servant standing at his side holding an umbrella over his head.

Some figureheads had odd extras – for instance, the female figure of Nanny the Witch on the front of the docked *Cutty Sark* is holding a horse tail made from real horsehair. The tradition here was that the horsehair plume was removed each time the *Cutty Sark* sailed, and then replaced in Nanny's hand each time the ship docked.

The Swill Basket Maker

Opposite: Britain's only swill basket maker Owen Jones at work on a basket that might be used for anything from logs to laundry. The swill basket remains unchanged in design in over three hundred years

Owen Jones is the only man in Britain still making oak swill baskets. He learned his trade at the workshop of one of the old school of swill basket makers: John Barker, who in his turn learned back in the 1920s when swill baskets were still common and made all over Cumbria.

Owen moved to Cumbria from Cornwall where he trained as an engineer, but a lifelong interest in basketry and crafts, combined with a family connection to John Barker, led to a partnership with the old man that lasted until John Barker's retirement: 'I learned the sort of things from John that you could never learn from a book,' says Owen, 'and if I hadn't met John and joined him in his workshop, the unbroken tradition of swill basket-making of which I am now a part might have ceased.'

The origins of the word 'swill' are, according to the *Oxford English Dictionary*, unknown. However, the earliest known reference comes from a will made in Richmond, Yorkshire, in 1596 where there is mention of 'six sand pokes with three great swilles'. By 1650 there are a number of references to swill baskets being used by millers, and in 1811 we know they were being used by washerwomen. Again according to the *OED* they were frequently used as a measure, particularly for oysters and herrings: a swill of herrings, for instance, was said to be somewhere between five and six hundred fish.

'They were certainly used as a measure in Cumbria,' says Owen. 'A 20in long swill basket was a "half peck", 22in long was called a "l'lle (meaning 'little') nick", 24in long a "peck" and 26in a "girt (meaning 'great') nick".'

The second half of the nineteenth century was probably the heyday of the swill basket: there were swill makers all over Cumbria and beyond because the baskets were used heavily in industry (by charcoal burners, for example), by farmers (for potatoes and broadcasting seed), and, of course, by ordinary people at home who might use them for anything from logs to laundry. Making a swill basket is a complicated business, but it all starts – for Owen, at least – with a visit to a small oak wood some two miles from his home at High Nibthwaite a few miles from Ulverston: 'I get all my oak from this wood. It's all coppiced, and I'm looking for lengths of timber about 6in in diameter. I use the first 8ft of the trunk. I then saw this into lengths of, say, 2ft to 5ft according to the final size I want each basket to be. Each length is then quartered, but not with a saw: to keep the strength of the wood, swill-basket timber is always cleft or riven – in other words, it is split. This means the wood will be naturally strong however thin the pieces, because you are following the fibres rather than slicing through them to create each piece. The end result is that each piece may rise and fall and be uneven where the fibres run round a knot.

'The quartered bits of oak trunk are then put into what we call a boiler, an iron trough filled with water, where the wood is boiled for a few hours. I then leave it all to cool down overnight before lighting the fire again and allowing the wood to simmer all day. Each billet –

Looking rather like a small woven coracle, the swill basket is made using a great deal of skill but only a few simple tools

Opposite: A well-crafted basket

that's what we call the quartered bits of trunk – is then riven while still hot: that just means I split each piece into increasingly thin strips. The ribs – we call them "spelks" – of the swill basket are usually about $^{1}/_{8}$in thick; the bigger bits of wood, or weavers – we call these the "taws" – are $^{1}/_{16}$in thick. There is an art to splitting the wood this thin – you can't just hit your riving knife and hope it will split the wood accurately: you have to do it in a controlled way.

'The basket is put together in what still seems a complicated way to me after all these years, but basically it starts with a steamed hazel rod, which is bent and fixed into an oval shape. I then smooth and shape the ribs using a drawknife and a "mare", or foot-operated vice. I can't easily explain the rest of the process, but it's basically a plain over-and-under weave starting from the bottom and working up. It's rather as if you started at the keel of a ship and then added the planks gradually up the sides.

'One billet might make twenty taws, which would be enough for two baskets, and it takes four hours to make a basket if you don't include all the soaking and steaming. The sizes range from 16in long to 36in long, which is a very big basket.'

A lot of locals buy Owen's baskets – they remember their parents having them, says Owen – and some go through a local ironmonger, but none to the wholesale trade: 'I like to be in touch with my customers,' says Owen, 'and swill basket-making is so labour-intensive that I can't do a wholesale price anyway. It would have to be low to allow for the retailer's mark-up, and then financially it wouldn't be worth me doing it.

'I've seen swill baskets that had been used in dry places that were fifty years old but, of course, a basket that is used in the wet will only last a few years, perhaps five or six at most; but they are wonderful things, and people get fond of them almost as if they'd found a friend.'

The Brick Maker

At one time only the very rich could afford to build with brick. Indeed, in Tudor and earlier times brick was a status symbol: strong, durable, weatherproof and attractive, it was used initially only for the houses of the great. New brick-built houses were often named, either officially or unofficially among the local populace, as Brick Place or Brick House, since the materials from which they were made gave them their unique identity.

Brick box and clay trimmer. At the top of the picture is a brick that has just been moulded. The indentations help the mortar hold the brick in position in a wall

The earliest bricks in Britain were made by the Romans – something not vastly dissimilar to a modern brick can be seen in the remnants of London's Roman wall near the Tower of London. However, with the arrival of the Saxons and other warlike Scandinavian and German tribes, a long tradition of timber building took over, and for nearly a thousand years after the Romans left Britain in the fourth century the art of building in brick and stone was more or less lost.

In the fifteenth century, however, bricks reappeared, probably at least partly in response to the increasing scarcity of timber. It is thought that the word 'brick' probably comes from the Old French *brique* meaning a fragment or a bit, and also a form of loaf. Since a loaf was baked, by extension it must have seemed perfectly obvious to transfer the word to the bit of clay that was also baked.

Right: One plain brick and two highly decorative brick styles suggest the enormous range that can be supplied by Coleford Bricks

In Britain the vast bulk of bricks have long been machine-made, but by rare good chance one company producing hand-made bricks survives: the Coleford Brick Company is based at two sites, one in Leicestershire and one in the Forest of Dean. The company has been making hand-made bricks since 1895, and is now run by the third generation of the Evans family who founded it. Mr Kerry Evans explains why there is still a demand for hand-made bricks:

'Well, today the demand is probably greater than it has been at any time in the past hundred years. Bodies such as the National Trust are always in need of the right sort of historic bricks for the structures they look after, and there are special one-off buildings that use hand-made bricks – for example, we supplied the bricks for the new British Library at Euston, London.

'Hand-made bricks are simple to make, although to do it quickly and well you need skill. You simply take the clay – a lump, or "clot" of it as the brick makers say – roll it in sand and then throw it into what we call a brick box, basically a wooden frame which will give the brick its shape. The standard modern brick is about 8½ by 4 by 2in; a Tudor brick would be nearer 10½in long by 2in deep. The forms for the bricks are wood or metal and they last about six months – as you can probably imagine, they are not given exactly gentle treatment by the brick makers! After the brick has been formed we dry it for three days and then fire it for four. We produce blue bricks – the ones used for patterns in many old walls – and all sorts of odd special bricks. Our bricks go all over the world.

'We employ eighty men on our two sites, and we're still a family-owned business, which is pretty remarkable in itself. We make about two million bricks a year, which is nothing compared to the numbers of factory-made bricks produced. But it is good that these things are still being made.'

Right: Here the 'clot' of clay has been thrown into the brick box and the excess clay is being trimmed off

Far right: Bricks by hand: Coleford make bricks for National Trust houses, for private restoration work and for special, highly prestigious projects – like the new British Library

The Thames Watermiller

The huge, ancient wooden mill wheel at Mapledurham in Berkshire still turns. Inside the mill, equally ancient wooden walls creak and groan, particularly when milling is in progress, as the building moves and settles. Like most ancient wooden buildings, Mapledurham Watermill is held together almost entirely by wooden pegs and joints. Throughout the mill are signs of the past: an old blackened candle-holder, scratched figures on the grain-dressing machine, and worn boards where generations of millers have turned in the same places on the wooden stairs. One miller has even carved his initials on the desk beside the window which faces upstream – here no doubt generations of millers sat and watched the barges bringing grain or carrying flour away.

Mapledurham Watermill is now run by Mildred Cookson, and it is the last working watermill on the Thames. Here you can see grain being milled as it would have been milled at the time of Domesday. But despite its antiquity, Mapledurham is a commercial concern, selling top-quality flour to local shops, farmers and bakers.

The Domesday Book records that 'William de Warene holds Mapledurham of the King ... there is a mill worth 20 shillings'. But even before William the Conqueror's great survey we know that there was a mill on the site because the Saxon chroniclers mention it. Much of the fabric of the present building dates from the fifteenth century when the mill served communities on either side of the Thames. Like

Opposite: Mildred Cookson standing outside a unique survivor – the Mapledurham Watermill. Its ancient, creaking structure remains substantially unchanged in centuries

Left: At the heart of the milling process is the grindstone. The surface of the stone is cut by a millwright who does the main furrows, and by Mildred who completes the tiny chisel cuts, or stitching

Right: Mildred – seen here checking the massive gear wheels – needs to be an expert on food production, engineering and weather prediction

Opposite: A giant gearwheel at Mapledurham: many of the teeth on this and other cogs and wheels were fitted more than a century and a half ago. When they next need replacing Mildred Cookson will do the work

most long-established businesses, it has known both prosperous and near-disastrous periods over the years. During the early 1660s, for example, the fact that the Plague raged in London brought the mill great prosperity, because the royal court – and that would have meant thousands of courtiers and attendants – moved from London to nearby Abingdon. Huge amounts of flour for the people and bran for their animals would have been supplied. And as London expanded in the eighteenth and nineteenth centuries the mill continued to do well, using barges to send its produce downstream to the capital.

By the beginning of the twentieth century, however, cheaper wheat flour from America and Canada spelled the end for the traditional British mill, whether wind- or water-powered. Hard Canadian wheat needed large electrically powered mills in place of stone-ground milling, and by 1920 the ten thousand or more working wind- and watermills in England were reduced to less than two hundred. On the Thames, where once there had been a mill every mile, they vanished – all, that is, except Mapledurham which struggled on, supplying flour and bran for the estate farms. However, by 1940 even Mapledurham could no longer compete, and the mill closed. But sheer good luck prevented it from being converted into a house or demolished. Then in the 1970s it was used as a film set, and this earned the mill enough money to pay for its own restoration – and by 1977 it was back in operation as a fully commercial business.

Mildred Cookson is clearly in love with Mapledurham. She began her career as a civil servant, but kept up her interest in watermills. By the time Mapledurham was restored she was well trained in the day-to-day running of mills, and when the estate owners, the Eystons, asked her to run the mill full time, she knew it was absolutely right for her.

Milling every Sunday and a few times during the week,

Mildred produces something in the region of a ton of flour a day; this is a tiny quantity compared to a big industrial flour mill, but what her flour lacks in quantity it certainly makes up for in quality. Mapledurham bread produced from stone-milled wholemeal flour '... tastes wonderful', says Mildred. 'It's because the stones work so slowly – they turn at 120 revs a minute compared with 750 in an electric mill, which keeps the natural oils in the flour to give it a lovely flavour.'

Left: Though it looks intricate and complex the mill's machinery simply allows the grain to be fed from a hopper at the top of the mill and then to gradually work its way down on to the grindstones

But Mildred's job is not restricted to milling: 'My job today is more or less like my predecessors' centuries ago, as I'm a Jack of all trades. I have to know some carpentry to make new teeth for the mill wheels – although some of the teeth on them have lasted 150 years, I have to understand engineering to know what to do if something is not working properly and I need to understand the mood of the river in case of flooding and high and low water.' Mildred has also taught herself to dress the millstones, a craft normally reserved for a millwright – of whom there are only about ten left in Britain. The 4½ft millstones at Mapledurham are French in origin and are extremely hard. They will last for upwards of fifty years – and are very expensive to replace.

The stones are cut or dressed in two ways: the main furrows Mildred leaves to the millwright; the tiny chisel cuts, or stitching, she makes herself (using what's known as a 'mill bill') in the flat areas between the larger grooves. This is where the fine grinding takes place, and the stitching has to be done with great precision because you need ten or twelve cuts to the inch.

Mildred's job is always packed with variety: 'I wouldn't change this job for any other,' she says proudly, 'particularly as I'm the only female watermiller in the country. You never know what will happen next – one day I'm showing a party of schoolchildren round, the next the river floods and I'm trying to prevent disaster… and the flour still has to be milled!'

The Somerset Hurdle Maker

Across the Somerset levels and beyond, willows were always grown in vast profusion to provide the raw material for thousands of basket weavers and hurdle makers. In recent years, however, the decline in these crafts and in willow-growing has been massive, and the ruins of withy plantations can be seen all over the county; this is particularly true in the Somerset levels, once the heart of the willow-growing and basket-making industry.

But despite the decline, willow-growing and the crafts associated with it are in a better condition now than they were in the 1950s when Stan Dare started making hurdles. By then plastics, the bane of most traditional crafts, had come in, and hurdle-making had almost vanished completely – at one time Stan was the only man in the area still making traditional hurdles. And yet in the 1920s and 1930s, hundreds of men were making Somerset willow hurdles to meet the huge nationwide demand from farmers.

'I reckon 80 per cent of the population in this area was once involved in willow-growing, basket-making or hurdle-making,' says Stan.

'But by the time I went from growing willows – which is what I did from a schoolboy – to making hurdles, the only people who wanted them were gardeners, who used them for fencing. The farmers dropped us completely.'

'If it hadn't been for Stan teaching his art to others, it would have died,' says Nigel Hector who runs the English

Opposite: Stan Dare at work: he grew willows as a schoolboy and has worked with them all his adult life. Speed and sureness of touch lie at the heart of willow hurdle-making

Left: A traditional hand-made willow hurdle

Hurdle Company at Stoke St Gregory in Somerset. 'Back in the late 1950s the art of willow hurdle-making was on the way out, and there was only Stan and a few others left working around the Somerset levels. Now we have a flourishing business, and it's all down to Stan who passed on his skills.'

The blackmole willow used by Stan is grown around Stoke St Gregory about nine miles from Taunton, where he still lives, and is harvested at any time between October and the end of March in pieces 3ft to 5ft long. To fashion it into a hurdle, the willow is intertwined by hand around uprights: 'It's speed and sureness of touch that's the art of making a good willow hurdle,' says Stan. 'My hands are not really up to it any more, at least not on a full-time basis, and I'll miss doing it. We always made our hurdles 6ft long and various heights, up to about 7ft.'

Between the wars most farms on the Somerset levels grew willow, but now the hurdle makers have to grow it themselves. Although in the last few years the industry has experienced something of a revival, it must be said that the glory days have gone for ever: 'All my family were involved,' says Stan wistfully. 'My brothers were basket makers. Originally we used only blackmole willow, but now all sorts of foreign willows are grown. They were harvested once a year, stacked and then boiled to make them pliable – they were boiled for two hours or thereabouts.

'A good willow hurdle will last about ten years, but often they last a lot longer than that. They went all over the country – I once made seventy-five for Diana Spencer's father, and that was just one order! My uncle managed eleven acres of willows all his life and he employed three men full time to help. But those days will never come back.'

Left and opposite: Willow hurdles have been made in Somerset for centuries: here Stan works the willow through the uprights from bottom to top

The Charcoal Burner

Before the Industrial Revolution, in the days when the Weald of Kent was one of the great industrial areas of Britain, charcoal was a vital ingredient in the manufacture of iron. The important thing about charcoal for the old ironmasters was that it burned at very high temperatures relative to wood, and thus made iron-smelting possible. Cannon for the Tudor and Elizabethan warships that sailed across the world were made in these now entirely rural areas because the abundance of hardwood meant plentiful supplies of charcoal.

To supply the ironmasters, large numbers of men were involved in the charcoal-burning business, but the history books have little to say about these shadowy figures who worked in remote locations and constantly moved about the country. If we know little about the charcoal burners themselves we do at least know something of the techniques they used to turn timber into charcoal. The main reason is that charcoal burning never really died out entirely in Britain. Though it has not been used in iron smelting for nearly two centuries, charcoal still had its uses until the 1960s and 1970s: until synthetic products made it redundant, it made an excellent filter for a number of important industrial processes and was also used to make artists' materials. Jack Durden, who was born in 1925, is one of the few men in Britain with a lifetime's experience of charcoal burning – he is thus the last link in a tradition that probably dates back at least to Roman times.

During a career that lasted more than forty years, Jack enjoyed a curious nomadic existence, constantly moving about the country from his native Dorset up through Wiltshire and Gloucestershire to the high beechwoods of Buckinghamshire. It all began after the war when Jack started work as a forester. He was originally employed to fell trees, but a chance meeting with two Polish refugees who had decided to stay on in Britain after the end of hostilities changed the course of his career:

'They were charcoal burners from a long line of charcoal burners – for them it was a family tradition, and I was fascinated by what they were doing. They had mobile kilns made of tin or some other light metal, and it seemed such an interesting occupation that I watched and learned and then set up on my own.' Soon Jack had his own kiln, and in those early days there was plenty of work. A lot of old woods had been neglected before and during the war, and the forestry people were happy to get him to come along and use up the mass of timber that the trade wouldn't take.

'The point is that you don't need a specific kind of timber to make charcoal,' he says. 'Almost anything will do, although of course if you were able to choose you'd always go for a hardwood. Beech is ideal, and oak probably the next choice. Softwood doesn't last as long and gives less heat, and as it's lighter it produces less charcoal. Since we sold it by weight, using softwood meant we earned less money.' When Jack retired in the late 1980s, a bag of char-

Opposite: The smoke from a charcoal burner's kiln conjures up an almost medieval image. Certainly charcoal making was a vital ingredient in cannon manufacture in Elizabethan and Tudor times, although today the craft is a rare one indeed

Overleaf left: The smoke-blackened features of the charcoal burner
Overleaf right: Loading timber into the base of the kiln. This has to be done carefully if the wood is to carbonise rather than simply burn

coal weighing about 20lb would probably bring in about £2.50.

Originally, charcoal burners worked by digging pits in which the burning took place, but this was labour-intensive given that, by its very nature, the charcoal-burning business had to move frequently to follow the tree fellers. Says Jack: 'The mobile kiln was ideal because it was light enough to carry. It came in two sections, each about 6ft in diameter, but the top about 4ft high and the bottom about 3ft.

'With charcoal burning the real skill comes in loading the kiln, especially the bottom half: you have to load it so the wood creates a channel for the air from the air inlets to the smoke outlets. The bottom of the kiln contains four smoke outlets and four air inlets – they're just holes really, but the wood has to be carefully stacked so there is a narrow gap from inlet to outlet. We used 4ft long pieces, rather like giant matches. On top of the stacked wood we'd then place plenty of what we call black wood – that's wood that hasn't quite become charcoal from a previous burning.

'In the top 3ft section of the kiln we stack more 4ft lengths – again like a box of matches, if you can imagine that – and then we put the lid on. Now if you imagine that the bottom section is 4ft high with 4ft lengths of wood arranged vertically, that's fine. But the 3ft section also has 4ft long pieces of wood in it arranged vertically – this means that the total height of the stacked wood is 8ft where the height of the kiln totals 7ft. This is very important, because when you put the lid on you want about a foot of wood sticking up above the top of the kiln itself. You put the lid on the wood and then light the black wood, which is now in the centre of the kiln, with an oily rag.

'With the four air inlets and the internal channel you get a tremendous draught and the fire will burn fiercely – it's a chimney on fire, really. Then the wood sticking out

Right: After the fire: Jack examines a kiln-full of top quality charcoal. The use of a mobile kiln enables the burner to follow the tree-fellers

Opposite: Jack shovels sand on to his kiln to seal it from the air. Without this – and it must be done at exactly the right moment – the timber within the kiln would simply burn away

of the top will gradually sink and settle; the lid will then go down and rest properly on top of the kiln; and you then seal round the lid with sand. It has to be a really good seal for the thing to work properly. The sudden lack of air in the kiln is what starts the carbonising process – that is, the process that turns the wood into charcoal. People think charcoal is just slowly burned wood, but it isn't – it's wood that's been carbonised at high temperatures.

'Once the lid's tightly on, there's no more through-draught and the heat can't escape so it goes back down into the stack of wood. So the heat builds up even more, and then, just when the smoke turns blue – and it takes experience to know precisely the right moment – you seal all the inlets and outlets at the bottom of the kiln. The whole thing is then left sealed for roughly forty-eight hours. If it is not properly sealed, the wood will all burn away and you'll have no charcoal. If all goes well, your 2½ tons of wood will have produced 7 hundredweight of charcoal.'

During his wandering years Jack lived almost continuously in a caravan which was moved from site to site. Towards the end he had three kilns continually on the go. 'You could certainly make a living at the charcoal burning,' he says now, 'but it became more difficult when charcoal filters were superseded by other products in the 1960s.

'We also used to make artists' charcoal and charcoal for pencils – you need willow for this, as its slightly pithy centre produces long, thin bits of charcoal that are not too brittle. But there was even less money in artists' charcoal than in the industrial stuff.'

But if there was little money in charcoal burning, there were compensations: 'The marvellous thing about it was that you got to live and work in beautiful remote locations away from noise and people – and that's a rare thing today.'

The Lobster-pot Man

Britain's west coast fishermen, from the far south-west of Cornwall to the extreme north of Scotland, have made a good living from lobster and crab fishing for at least two hundred years. Anywhere where the seabed is rocky is likely to harbour that most sought-after crustacean, the lobster.

Today, sadly, lobster and crab are under enormous pressure from overfishing: the economies of scale applied to commercial fishing mean that bigger and bigger ships are used, together with increasingly sophisticated fish location techniques and increasingly efficient lobster pots – and more of them. But the single-handed fisherman – the man who makes his own pots and fishes in a way that is inherently sustainable because it does not over-exploit the resource – is still with us, though, of course, he is now something of a threatened species himself.

One such is Nigel Legge of Cadgwith near the Lizard in Cornwall. Now in his forties, Nigel has been lobster fishing since the day he left school. He started with his father and learned the trade from him. He now owns his own 18ft boat and fishes alone right through the season. For lobsters the season is April to mid-August, at which time he turns his attentions to the crab, the season for which lasts until November. Then in mid-November the real work begins, because unlike almost every other lobster fisherman in Britain today, Nigel still makes the old willow lobster pot. Almost everyone else has gone over to plastic and steel pots, but Nigel will have none of it:

'There's something marvellous about the old pots – partly I think it has to do with the skill of making them. It's difficult, and you need to learn it as a youngster as I did from my father at a time when all the fishermen made their own pots. You had no choice then. It was only in the mid-1960s that alternatives became available, and from then on, of course, the traditional lobster pot was more or less doomed. But apart from the pleasure of making the old pots, there's something satisfying about catching your lobsters in something you made yourself.'

Opposite: Nigel Legge spends the winter making traditional lobster pots

The willow has to be soaked for several days until it is pliable

Nigel has that look of enormous strength and resilience that seems to be the physical hallmark of the commercial fisherman, but his sense of the importance of tradition also makes him unique. But how do you make a traditional lobster pot?

'Well, it takes about four hours to make a full-size one – that's about 3ft across the bottom and about 18in high. It's really a basket-weaving skill combined with a few fishermen's tricks – like leaving the willow soaking in the sea for a few days to make it pliable enough to work. I get my willow from a Somerset basket-maker – willow from the Somerset levels is definitely best. The pot has to have a

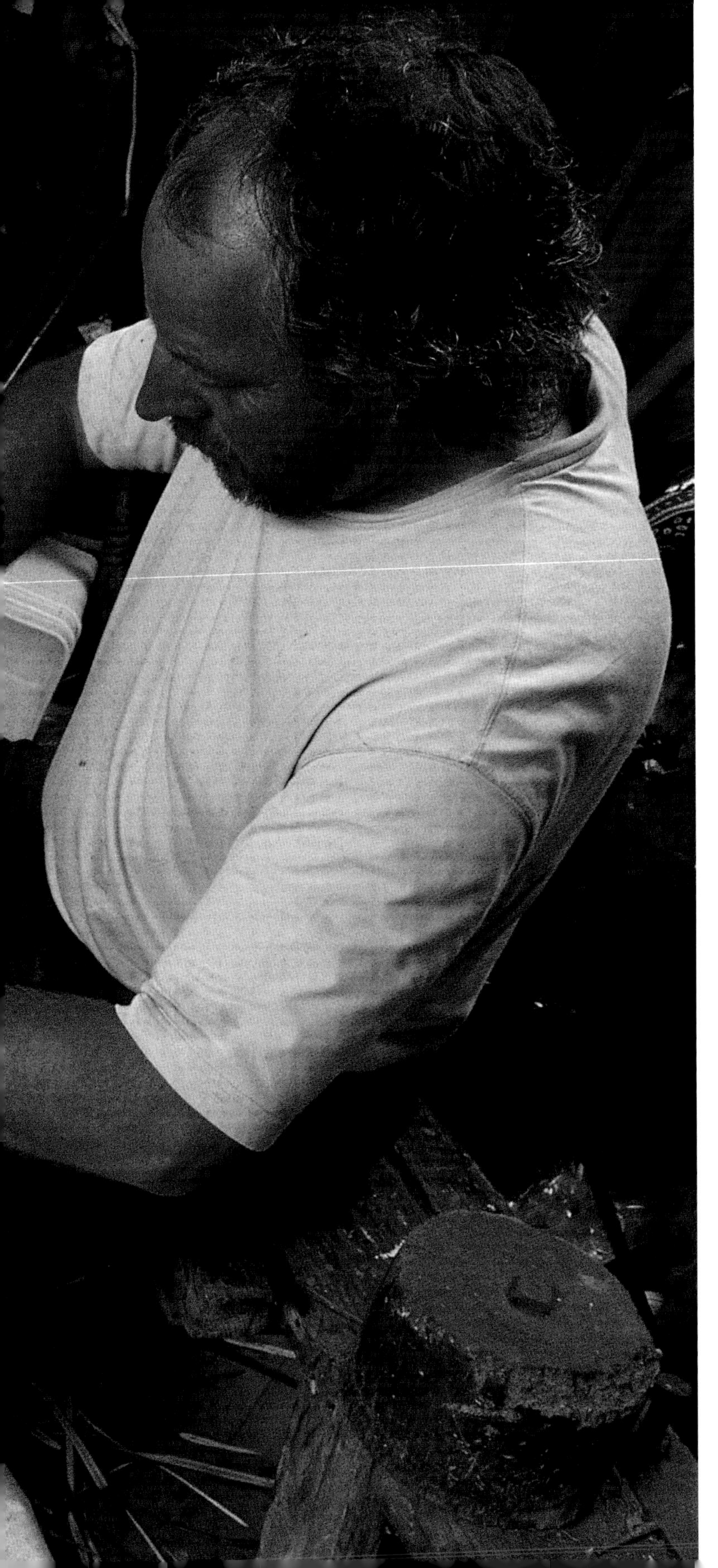

carefully woven funnel through which the lobster or crab can make his way to the bait. It has to be just right so he can get in, but not back out again. But lobsters are crafty creatures, and if you leave your pot down too long they will find their way out. A good pot – and all mine are good 'uns! – will last for several years.

'Sometimes I use local Cornish black willow withies for the bottom of the pot because they are much tougher than the Somerset kind, but the whole pot is made from willow of one kind or another.'

Working his 18ft boat single-handed out from Cadgwith, Nigel would expect to fish anywhere from right under the cliffs to about one mile out. In eight hours he can set 180 pots which, two days later, he will spend another eight hours hauling out of the water. It's arduous work, but highly satisfying when he gets a good haul – particularly as he'll have used his own pots rather than pots made in some anonymous factory.

'I don't really like the modern ones – it's not very satisfying using them, and they're no more efficient than the old ones. Besides, I keep my pot-making for the winter months when the fishing stops anyway. If I didn't make the pots – and I really enjoy making them, apart from anything else – what else would I do?' The enormous appeal of the traditional lobster pot can be judged by the fact that Nigel recently received an order for several hundred from Laura Ashley, the clothing firm.

'That kept me busy for a while, I can tell you,' he says with a chuckle, 'but I get lots of smaller orders from people who just want to keep flowers in them. I'm always happy to oblige.'

Left: Nigel weaves the willow through the upright canes

The Irish Bagpipe Maker

Opposite: Eugene at work on what is by all accounts one of the world's most complex musical instruments. The Irish, or 'elbow', pipes are fiendishly difficult to play

Below: Chanters and drones: the constituent parts of the Irish pipes are each made up of a number of intricate parts

According to Eugene Lambe, the Irish bagpipe is the most highly developed in the world. In fact it is so highly developed and complicated that even working full time, Eugene can only turn out about a dozen in a year: 'The thing is that Irish bagpipes cannot be mass-produced, unlike Scottish bagpipes; they are just too complex and intricate, and one of the reasons for this is that they are the only bagpipes in the world that can play two octaves.'

Despite being made originally as an instrument for the masses – basically for the poor – the pipes were always made with great skill and with the finest materials: 'It's a myth to think that the old pipe makers only used local woods,' says Eugene, whose workshop is in Galway in the far west of Ireland. 'Like pipe makers today, they imported the finest timbers to make their pipes. African black wood, the sort used to make oboes and clarinets, is used for all the acoustic parts of the instrument – the chanters and the drones – and it's been used for at least two hundred years, together with Spanish reed, and brass for the metal parts.'

The only concession to the modern world among the handful of pipe makers still in business today is the artificial leather used for the bag. Nonetheless, tradition is a central part of what makes the pipes so fascinating, although the history of the instrument is anything but straightforward, as Eugene explains:

'Well, the Irish pipes are not actually all that old – they probably date back to the 1700s. Before that time the Irish bagpipe was more like the Scottish pipes we are all familiar with today, but that sort of pipe is designed to be played outdoors and, for political reasons, as I shall explain, the Irish pipes had to change: namely, in the seventeenth and eighteenth centuries Irish culture was seen as seditious by the government – which, of course, was British then – so they tried to suppress it. Many so-called seditious pipers were hanged simply for playing their pipes, so the music was forced indoors, even underground, and this is when

FIRST

the Irish pipes we know today came into existence. They were smaller and quieter than the pipes that had gone before, and were designed to be played inside people's houses, quietly and privately – the bigger pipes have been extinct since that time, of course.'

The Irish pipes are also called the union pipes or the 'uilleann' (Gaelic for 'elbow') pipes: 'There are several theories about why they are called the union pipes,' says Eugene. 'Some say it is because they were very popular at the time of the Act of Union in 1801 between Ireland and England; others say that the pipes got their name from the fact that they united drones and chanters.

'Until very recently most people learned to play by ear, and the older players still know off by heart as many as five or six hundred tunes. Visitors are always amazed that Irish players, or at least the older ones, cannot read music at all – the tunes are all in their heads.

'In the 1950s only one pipe maker was still going in the whole of Ireland, and the pipes nearly died out altogether. There was an idea then that Irish music was somehow backward. Then, gradually, traditional music began to come out of the doldrums. The Clancy Brothers started to make it popular, and now the pipes are probably more popular than they've ever been. I have a three and a half year waiting list for my pipes, and they are expensive – probably more expensive than a set of any other pipes – about £2,500. But then there are seven reeds, the chanters,

Left: Making the reeds: each set of pipes has seven reeds and like every other part of the instrument, they are made, painstakingly, by hand; (top) Eugene uses a chisel to shave a reed down to size, while (below) the reed is fitted to one of the chanters

Opposite: Using experience and a keen eye, Eugene puts the finishing touches to a reed

the drones, a strap to fix the bag – which carries the air – round your waist, the bellows, and the mechanism that is opened and closed by the player's leg. The Irish pipes, you have to remember, are played sitting down.

'It is enormously difficult to explain how the pipes are made because so many complicated pieces have to be fitted together – regulators, keywork, leatherwork and woodwork, reeds and so on. Most of the old makers were players first and makers second – the man who taught me was one of the last of the old school. He made my first set of pipes on a lathe in his back kitchen, and I was so fascinated by watching him that I learned to play and then began to make them myself.

'It used to be said – and it wasn't far from the truth – that it took twenty-one years to learn to play the Uilleann pipes. That's a bit of an exaggeration, but they are enormously difficult because there are so many different things to co-ordinate. The keys are under your right hand, the bellows are played with your left, and the chanter rests on your knee. To play the second octave the player silences the chanter on his or her knee and this doubles the pressure on the reed; no other pipe can do this. The basic technique is a smooth in-and-out with the elbow that pushes the air through the drones (the pipes without holes) and through the chanter (the pipe that has finger holes).

'At the heart of the pipe-maker's skills is the lathe on which the chanters and drones – the pipes – are turned.'

Eugene now plays his pipes all over the world, as well as making them, and he is optimistic about the future for this fascinating instrument: 'There are Irish pipe players in Tokyo now, and all over America where Irish pipe-playing has a continuous tradition dating back to the famine years of the 1850s and 1860s when millions of Irish men and women left for America.'

Opposite: Eugene makes only a dozen or so sets of pipes each year. This has a great deal to do with their complexity, but Irish pipe playing is popular as far afield as Japan and America

Left: Eugene with a recently completed set of pipes. 'It's a myth,' he says, 'to think that the old pipe-makers used only local woods'

The Freeminer

Gerald Haynes is one of the last of the freeminers of the Forest of Dean. Traditionally seen as something of a secret world, the Forest of Dean remains one of those curiously unexplored parts of Britain. In former times it was an area rich in industrial workings: limekilns were common, and charcoal burners and iron smelters would be found in most of the woodland areas, along with coppicers and chair makers. Most of these trades died out long ago, and the last big commercial mine vanished more than thirty years ago. But here and there in quiet corners, men like Gerald Haynes continue a tradition of mining that probably dates back to the Iron Age.

The freeminers are a unique group of individuals whose status and practices are without parallel in the UK – and according to Michael Johns, who is their chairman, they are fiercely protective of their rights. The story of how they obtained these rights in the first place is a fascinating one.

During the English/Scots wars in the fourteenth century, Berwick-upon-Tweed was besieged on several occasions. At one of these sieges the situation had reached a two-year stalemate: no progress could be made by the English against the Scots who held the town and castle. Then Edward II, or one of his military advisers, had the idea of using sappers – miners – to dig tunnels under the walls of the castle. The king sent for the most famous miners in the kingdom, the men who worked the mines in the Forest of Dean; and when they had done their work, the English entered the castle and Berwick

Opposite: Coming up for air: freeminers like Gerald Haynes are a breed apart. They work alone underground as part of a tradition that dates back to fourteenth-century Berwick-upon-Tweed

Left: Gerald Haynes; the freeminers of the Forest of Dean enjoy unique privileges that date back centuries

Opposite: The seams worked by freeminers are so narrow they often have to be dug out by hand. Here Gerald uses a pick to dig out the coal at the face

was won. Those fourteenth-century miners came from a very specific area within the Forest, and the king was so pleased with what they'd done that he decreed that any man born where they had been born – that is, within the ancient boundary of the Forest of Dean and the hundred of St Briavel's – was entitled to mine for coal there in perpetuity. To qualify as a 'free' miner, each man had also to have worked at the coalface for a year and a day.

Since that time freeminers, as they became known, have carved a precarious living from the tunnel-ridden ground; but as the price of coal dropped in the 1960s and 1970s, freemining became increasingly difficult and uneconomic; today only Gerald Haynes makes a full-time living from it.

'It was never a job that was going to make you rich,' says Donald Johns, who spent over forty years mining. 'And most young people today don't want a job that involves long hours of back-breaking work for little reward. I always enjoyed it, despite the hard work – there is something very different about being a freeminer compared to being a miner for British Coal, for example. You are your own boss, and you work your own seam in your own time. Mind you, that has its own hazards. You might get a few tons of coal out of your mine every day for weeks, and then a rockfall will stop you working for days or even weeks. The freeminer does all the shoring up, fixes the rails for his trucks, pushes or winches the filled trucks out of the mine and hacks away at the coalface. Most of the work is still done with pick and shovel as the mines are not generally big enough for mechanical cutting gear, and the miners couldn't afford it anyway.'

As Donald explains, it is still relatively easy to get your own mine – just so long as you were born in the right place: 'You can inherit a mine, and some do, but you can also take out what's called a "gale". That's what the Forest mining area is divided up into. You present yourself at the deputy gabeller's office – he's the keeper of the coal, he regulates the mining. He has a book where miners have registered for centuries. If there is a spare gale and the applicant meets the requirements of a freeminer, he pays a fee and off he goes.'

Donald originally worked his mine with his two brothers, but in more recent times he worked alone: 'It is dangerous working alone, but I've always put safety first, and I've hardly ever even scratched myself. Forest coal is good coal – we used to sell most of ours to the power station in South Wales. The biggest problem here was faults and flooding; that's why the big commercial pits closed.'

Freeminers need many skills. You have to be an engineer and a collier, fitting wire ropes and pulleys to winches, adapting gears and couplings to various engines to haul out trucks when the gradient is too steep for a man to walk. Says Donald, 'We had no giant machines to get the coal out, and we don't have shafts here – our pit entrances are really just paths on an incline, and the shaft slopes gradually into the ground. The technique is to load your truck and then winch it up the slope. In the old days they'd have used horses, or if the incline was shallow enough a man might push the truck up the rail to the surface himself. But it was heavy work because a truck might have six or more hundredweight of coal in it.'

Forest of Dean mines have curious, sometimes ancient names, such as Random Shot, All Profit, Speculation, Favourite, Arthur's Folly and many more.

'The vital point about freemining,' says Donald Johns, 'is that it will never entirely stop – it may even blossom again when cheap foreign imports of coal dry up or become too expensive. That would make freemining far more economic, and the reserves of coal that remain would last centuries. While there's coal there will always be freeminers, perhaps only one or two, and they'll never quite give up. King Edward said they could mine forever, and mine forever they will!'

The Wooden Oar Maker

Leon Pezzack makes wooden oars, paddles and spars. He doesn't know if anyone else still makes them, but he is certain that 99 per cent of oars made today are either plastic or carbonfibre. His workshop is a small room above a barn with views out over the Cornish coast near Mousehole, and he is the first to admit that it is the Cornish love of traditional boats that has kept traditional oar-making alive:

'We are one of the last places where traditional wooden boats with fixed seats are used for rowing competitions; it also happens that we're the best in the world at fixed-seat rowing, and one of the rules of the sport is that in these wooden boats you have to use traditional – and that means wooden – oars.' Plastic and carbonfibre are comparable in price, says Leon – but Leon's wooden oars are built to last, and he can make them to individual specifications. Each oar is hand-made in a slow process that involves some seven separate pieces of wood being fitted together.

'There's a steady demand for these oars,' says Leon. 'I've never had an order for anywhere beyond Bristol, but in Cornwall the popularity of rowing in wooden boats has increased so much in recent years – I reckon there's about a thousand traditional rowers in the county – that there's plenty of work for one man.'

Leon is chairman of the Cornish Rowing Association, and a fierce advocate of the sport: 'Our 15ft boats have three men rowing; an 18ft craft, called a flash boat, is rowed by four people and coxed using a tiller and rudder. You've got to be strong to row these boats, but we help them a little by making our oars hollow. In Cornwall there are also six-oared Cornish pilot gigs that were also used in pilotage of sailing ships. These were used when the distance travelled was further than usual, or the port of call more exposed.

Opposite: Leon Pezzack in his tiny cluttered workshop which overlooks the sea near Mousehole

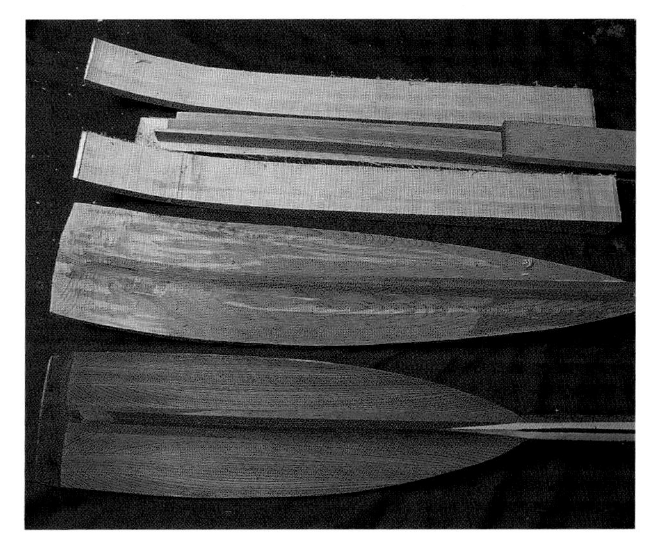

How the blade is made: (top) the sections before fitting and glueing; (centre) the blade glued and shaped. The finished blade (bottom) is smooth as glass and perfectly shaped

'My oars are always made from spruce with an oak or elm tip and a redwood blade. Elm is probably best for the tip because it has such a tangled grain that it won't split. Spruce is wonderfully light, strong and stiff, and this is just what you need for rowing.' Using a simple set of very old tools – draw knives, spoke shaves and planes – Leon makes just one hundred oars a year.

'The three classes of traditional fixed-seat rowing boat used in Cornwall today all have regulations restricting their construction to traditional materials. This also applies to oars and paddles which vary in length from 9ft 6in to 13ft.

'We also make masts and spars for traditional sailing boats like gaffers and luggers.'

The Trug Maker

When the willow boards have been nailed into position on the trug, Tim carefully trims the ends with a knife

Opposite: Tim Franks completes another trug in the old truggery at Herstmonceux. Though some manufacturers now use plywood, Tim sticks to the old way using sweet chestnut and willow

The word 'trug' comes from the Anglo-Saxon word trog meaning a wooden vessel, trough or boat, and though trugs have been made in Britain for centuries, they only really came to the attention of the public after the Great Exhibition of 1851. While visiting the exhibition Queen Victoria ordered several trugs of the type displayed by Thomas Smith, one of the great Sussex trug makers. Smith was so pleased that he apparently took the Queen's trugs to London himself in a wheelbarrow!

Once it was known that Victoria was using trugs, everyone wanted one; gardeners all over the country started to use them, and the trug-making industry blossomed in Sussex. Directly and indirectly, trug-making provided thousands of jobs for the next century or more in the Herstmonceux area of Sussex, always the centre of the trug-making industry. Then in the 1960s came disaster, as increasingly plastics made wooden buckets, bowls and baskets a thing of the past: trug-making nearly died.

Only a handful of traditional trug makers survived the really hard times, but survive they did, and in recent times there has been a resurgence of interest in this ancient art. True, there are some manufacturers who use plywood to mass-produce trugs that have a vaguely traditional look about them – but the old sweet chestnut and willow trug is now a rare animal. One of the few genuine makers is Tim Franks who works alone in The Truggery in the hamlet of Coopers Croft near Herstmonceux. Tim has been making trugs all his working life, and his workshop has been used continuously for trug-making since 1899. A 'Truggery trug' is still made to a design that hasn't changed since the workshop opened.

The old Sussex trug comes in sizes numbered 1 to 10. The smallest, size 1, measures just 7in long and has a capacity of one pint; the biggest, at one bushel, is 30in long and can carry a sack of potatoes. According to Tim, each trug size has its adherents:

'Well, professional gardeners tend to go for the half-bushel size, a size 8, but the most popular in terms of numbers sold is the size 7, which takes 2½ gallons and is 20in long. Between sizes 7 and 8 there is a distinct change in the way in which the trug is made. Up to size 7 a trug has feet made from willow; at size 8 and above, trugs are made with chestnut straps, two forming a V-shape running underneath. Occasionally we get orders for special sizes – I once made one the size of a chicken's egg, and there's no reason why, in theory, you couldn't make one 10ft long!'

Tim is modest about the manufacturing process, insisting that it's all quite quick and straightforward; but the real work is in the preparation, and only the skill of a man who has spent all his working life making trugs could make it all look so effortless: 'From start to finish a trug takes about two hours, but if you had all the pieces ready prepared the actual assembly of the trug would take only about ten minutes. The real skill is cleaving and shaping the various pieces of wood before you put it together.'

The willow used for the trug boards – the flat slats from

Opposite: Essential ingredients: the completed trug (right) contains a drawknife, nails, pincers and a hammer. On the left the shaved willow boards and hoop of chestnut ready for assembly

Below: Tim uses a drawknife to produce a thin, but very strong, trug board

which the body of the trug is made – comes from a local cricket-bat willow-grower, and the sweet chestnut, which is used for the hoops and handles, is also available locally. Indeed, trug-making probably grew up in this part of Sussex precisely because there was such an abundance of chestnut. After the willow arrives at the workshop, Tim cuts and shapes it into boards which are then soaked in a barrel of rainwater for half an hour or so.

'I know it sounds a bit unscientific, but that's how it's always been done,' says Tim with a smile. 'By the time it's ready to be nailed into place, each board is about ⅛in thick. The sweet chestnut for the hoops and handles is cut in winter and brought straight into the workshop to be cleaved – that is, split rather than cut. This helps it keep its strength because

it's the hoop that keeps the trug together. The chestnut hoop is steamed and shaped around a block before being nailed. The chestnut we use can't be from a branch, it has to be from the main trunk of the tree to be of any use. Once you have the hoop you begin to nail the boards into place. The widest board is nailed into place first; this is the one that makes the bottom of the trug, as it were. After that the other boards are progressively nailed on, getting narrower as they go up the sides.

'All the tools we use are simple and ancient: a drawknife, a horse which is a sort of foot-operated vice, an axe for cleaving, and a hammer and nails.

'The great thing about a trug is that as it gets older, it gets tighter – in other words it gets stronger, and even if you mistreat it, a trug will last ten or fifteen years. We've had fifty- and sixty-year-old trugs in for repair, and with a little work they can be made good as new.'

Tim makes a remarkable range of trugs, too: 'The point is that trugs were traditionally made in a range of styles to do a range of different jobs; we therefore still produce oval trugs, round ones, square ones – very popular with Americans – flat ones, and cucumber ones which are long and narrow. We even make a walking-stick trug, a stick with a trug positioned two-thirds of the way up it which was used by ladies to carry their sewing. Then there's the stable basket, big and round with no handles. As the name implies, it's used to feed horses, although we sent one to Prince Charles at Highgrove House which was to be used as a dog basket.

'We get lots of orders from people who collect trugs, and the grand-daughter of the founder of this business – a lovely woman called Molly Dean, who is now in her nineties – comes to see us regularly; she's always got lots of advice for us if we're not doing things properly! It's a real link with the old days.'

The Balloon Basket Maker

Aubrey Hill began making willow baskets with his father, a trained basket maker, more than fifty years ago. In the early days the two men made pigeon baskets almost exclusively. After the war British Rail still carried livestock, and pigeon baskets were much in demand, but when that source of business dried up when British Rail stopped carrying livestock, Aubrey decided to make picnic baskets.

'We were lucky,' he says now, 'because we were able to adapt, and when one kind of basket was no longer wanted we found another that was. That's how we survived the bad old days when plastics came in and people wanted everything in plastic just because it was the new thing. The good thing is that it didn't take them long to realise that plastic is not the be all and end all.'

The biggest change to the Hills' traditional basket-making business came twenty years ago, and it came right out of the blue: 'We were busily making picnic baskets when the phone rang and a chap asked if we could make really big baskets. I said we could, even though we hadn't made any at that stage, and the next day he turned up with an old balloon basket on the top of his car. He wanted a replacement, so we made it for him, and since then the orders have just continued to flow in. I should think that making balloon baskets now represents about a third of our turnover.

'We still use local willow for many balloon baskets, but these days some people opt for baskets made from cane, in fact an Indonesian cane known as "kooboo". Willow comes in perhaps 8ft long pieces that taper, but kooboo cane grows in straight pieces up to 100ft long. We use kooboo for the uprights anyway, for all balloon baskets. When people say to us, "What's this kooboo like, then?" I always ask them to remember the thing that looked like a creeper that Tarzan used to swing on – that's kooboo cane. Kooboo produces a butter-coloured basket, whereas willow is a sort of reddy-brown.'

Opposite: Father and son: Aubrey and Darrell Hill – 'We made the world record basket, a double-decker that would hold fifty people'

The tools of the trade could hardly be simpler: (from l to r) a sharp knife; a bodkin; an iron; and a pair of secateurs. The techniques used for big-basket making differ little from those used for picnic hampers and shopping baskets

The process of making a balloon basket is pretty much the same as it would be for any other basket, it's just a lot bigger. In fact Aubrey, who now works with his son Darrell, has made some of the biggest balloon baskets ever made: 'We were asked to make a basket 11½ft long by 3ft high – that size would easily carry fifteen to eighteen people. On another occasion we made the world record basket, a double-decker that would hold fifty people. So you can see, willow and cane baskets are very strong.'

It takes two or three weeks to make a balloon basket, depending on the size, and as ballooning has increased in popularity, so has the demand for willow baskets. But surely modern materials will out-perform willow?

'That's what everyone thinks, but the answer is "No".

They've tried all sorts of space-age materials for balloon baskets, but nothing works quite as well as willow. They tried kevlar and it just broke. The thing is, when a balloon comes down there's a lot of bumping and scraping, yet despite these stresses and strains the willow basket quickly returns to its original shape. So it's strong and it's flexible.

'Ballooning is now incredibly popular; there are companies which organise champagne balloon flights for people's birthdays and anniversaries and so on, and there are thousands of private balloonists all over the world – these are the people we supply.'

In addition to balloon baskets, Aubrey and his team of twelve – which includes his son – make cane furniture, baskets of every description – and even coffins! But as Aubrey observes: 'Well, it isn't as strange as it sounds, in fact, because traditionally in this part of the world people were buried in basket coffins. I had one lady ring us up and order her own coffin – she even gave us her measurements! People like the idea of them because willow is environmentally friendly.'

The basic tools of basket-making are the same, whether one is making a balloon basket or a smaller one: a pair of secateurs to cut the cane and willow; an iron, which is a tool to tap the woven willow down; and a bodkin to push between the weave. The real skill is in the weaving, and for that one needs quick, dextrous, experienced hands.

A balloon basket will last for many years, regardless of how many hard landings it has to put up with. In fact says Aubrey with a grin: 'It's ironic, but willow baskets are now ousting plastics! No one wants it any more, they all want willow or cane.'

Left and opposite: Despite the invention of numerous high performance materials – like kevlar – wicker baskets are still the best for ballooning. As this example, nearing completion, shows, they look good, too!

The Flint Knapper

According to Phil Harding, flint knapping is the world's oldest craft. Originally trained as an archaeologist – one of the few jobs that allows him to exercise his flint-knapping skills – Phil happily admits that he is obsessed by flint and all its uses: 'Knapping is the process of working flint,' he explains with enormous enthusiasm. 'You don't cut flint, you knock or push flakes off it, but it's such a wonderfully useful material. It's pure silica, brittle and yet strong. For centuries it's been used as a building material anywhere and everywhere it is found.'

Norfolk is the great centre for flints, and in prehistoric times flints were mined at Brandon (actually just over the county border in Suffolk); here visitors can still visit Stone Age flint mines at Grimes Graves. In the seventeenth, eighteenth and nineteenth centuries the same area supported an important industry of gun flint makers, the sort of flints used in old muskets and pistols to provide a spark for the powder: 'You could say,' says Phil with a grin, 'that the defeat of Napoleon at Waterloo was entirely the result of the flint knapping that took place at Brandon. Without the flint knappers the main weapon of warfare of the time would not have existed.' Today flints are still worked for guns used by historical societies such as the Sealed Knot, an organisation which re-enacts old battles.

'Dozens of men worked at Brandon making gun flints,' says Phil, 'and it's a highly skilled business. They made blades

Opposite: Phil Harding examines a flint cut into a brick shape: knapping flint with right angles takes great skill

Left: Phil uses a 'punch' to knap a Norfolk flint. Prehistoric man would have done the same job using deer antler

– that is, long strips of flint from which three or four individual gun flints could be taken. And as they made each flint they would know by just looking at it whether it would do best in a musket, a pistol or whatever.'

Phil has made many gun flints. He also makes replica flint tools as part of his job as an archaeologist: 'My work has included making Stone Age flint tools, axe heads and arrow heads for testing by various researchers and archaeologists; however, most knapped flints are today used for building work. I remember I did about a thousand for West Dean College in Sussex. Each quoin – a sort of corner piece for a building – had to be just right, and it's not easy work. I think I managed to do about a dozen a day, which is slow going when you've got a thousand to do!

'Whatever you are planning to do with flint, half the battle is choosing your flint in the first place. The best-quality flint comes from East Anglia, but I've had bad flint from East Anglia and good flint from other areas where the flint was said to be consistently poor. Poor flint is flint that has been exposed to excessive cold or excessive heat. That doesn't mean that a piece of flint will be poor just because it's been left out on one frosty December night; when we say a piece of flint is "frosted" we mean that it was probably exposed to extreme cold during the last Ice Age about ten thousand years ago – it would have been damaged if it was lying on or near the surface during the glaciation.

'Flints left in heaps in the summer sun also deteriorate because despite the fact that they are hard, they do contain moisture, and if they dry out they become difficult to work – that means they won't knap or cut crisply. Frosting on the other hand creates fracturing, and a badly frosted flint will probably shatter as you try to work it.'

Wherever you find chalk you find flint, so there are rich pickings right across the South Downs through Sussex, Hampshire, Wiltshire and Dorset. Flint also occurs through the North Downs and up through the Chilterns and east to Norfolk. It peters out at Flamborough Head.

'In parts of Norfolk, flints were often used whole,' says Phil. 'The builders would simply take a wagon to the beach, load it up with fist-size spheres of flint – these would have been washed into shape over centuries by wave action – and then simply build them into the mortar walls. The flint wasn't there just to fill the wall, it was what gave the wall its strength. With bigger pieces of flint the idea is to create one flat surface – that's the surface you would see – and the irregular shape of the rest of the flint would be used to lock the flint into another flint within the wall. A flint that can be used like this is said to have a good "tail". The classic example would be a concave piece of flint being bedded into a convex piece of flint.

'When you take an old flint wall down you often see how the builder has linked much of the flint together like this. The irregular shape of the flint within the mortar is obviously a good idea, as it grips other bits of flint and the mortar. Mortar for flint work has to be soft and dry – soft because the flint will move, shrinking in cold weather and swelling in hot; and not too wet because, unlike brick, flint won't absorb much moisture and wet mortar will simply ooze out of the wall.

'Only in rare instances will a flint be made to look like a brick. I have seen this sort of thing, but it's very difficult to do because you can't easily knap a flint at 90 degrees – 70 or 80 degrees is usually your maximum. And, of course, to make a brick, you're knapping at, or very near, 90 degrees.

'The greatest flint knappers were probably the prehistoric Danish people and the Egyptians – but I reckon I could match the work of the best British prehistoric flint workers! The technique for flint knapping varies. Ancient man used deer antler to knock flakes off the flint, and this works because the antler is softer than the flint so it is less likely to shatter it – and it doesn't matter that it's softer because you are working at angles of less than 90 degrees and the flint is very weak at these angles. In fact we still use deer antler today, though equally we'll use what we call a "punch", a sort of blunt chisel which is hit with a mallet; but you can also use a stone, or flint on flint.

'The real skill is being able to visualise how the flint will look when it's finished before you've started. What you have to remember is that no two pieces of flint are the same, because when you approach each one you have to tackle it in a different way. You assess the angles before you start, because taking off a series of flakes of flint can easily wreck the whole piece if you don't know what you're doing.'

Knapping means removing flakes of flint in a process that is almost as old as mankind. 'It's the world's oldest craft,' says Phil

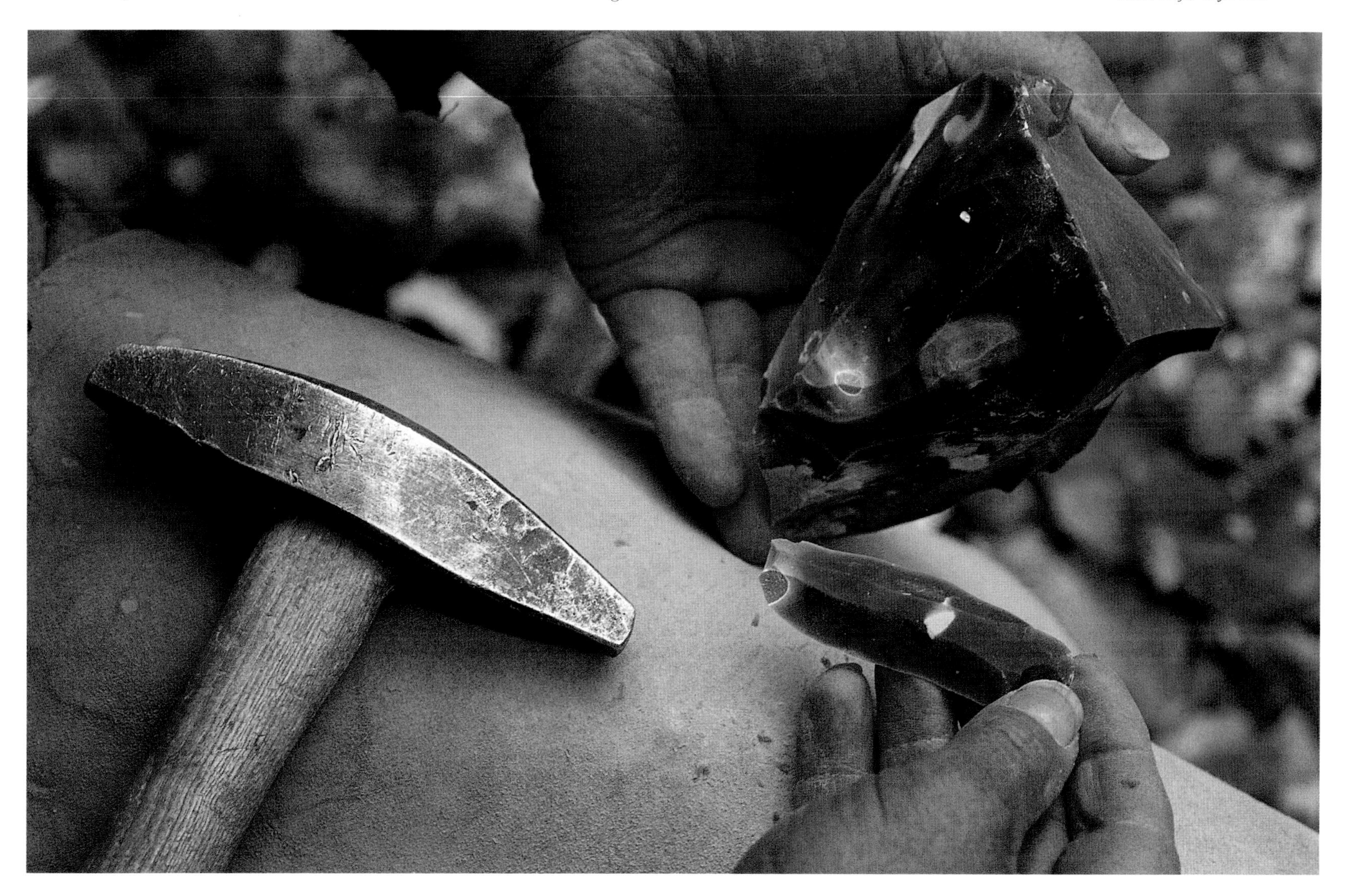

The Reed Cutter

Opposite: *Reed cutting in the tranquil Norfolk wetlands*

Below: *Marsh man, Eric Edwards*

'It's the best roofing thatch there is,' says Eric Edwards, one of the last of the Norfolk reed cutters. Sixty years ago vast areas of Norfolk's wetlands produced a rich annual crop of reed which was sent to thatchers all over the country. Then thatch began to be replaced by tiles and slates and the old reed cutters gradually died out. When Eric started on the Turf Fen Marshes near Ludham (where he still works today) the craft of reed cutting was already in decline, although plenty of the old hands were still around to pass on their skills.

'There's actually more demand for reed now than there was when I started as a young man in 1967,' he says with a smile. 'The point is that during the 1980s and early 1990s people began to realise that the countryside would lose much of its charm without a few thatched houses. New people learned the old trade of thatching and it became something of a boom industry. I think people also realised that thatch was actually a very good roofing material. It's very warm – a superb insulator – and Norfolk reed thatch will last as much as eighty years.'

Today, despite the renewed popularity of thatch, only a

few reed cutters remain. Much of the market has been lost to cheaper reed from abroad – from Turkey and Poland. Some mechanisation has come in, too, among the few remaining Norfolk reed cutters, but Eric still works occasionally by hand.

'Most reed is cut mechanically, but there's a real skill in mowing reed – that is cutting it with a scythe. I make my own scythes – you have to since you can't buy them any more because no one makes them.'

Eric explains that the reed cutter's scythe is much straighter than that used for hay cutting. 'I just work my way along a dyke, find an alder growing naturally with the right sort of bend, cut it, add some handles and a blade and there you are.

'A scythe is still much better than the mechanical cutter when the water is high so it's not just sentiment that keeps me doing it the old way. I'm also proud of the fact that I'm one of the few cutters – possibly the last – who know how to mow reed by hand in this way. The only other tool we use is the rake – not like a garden rake, but a long pole with three nails in it. The rake is used to clean the reed – that is strip off any roots and bits of side shoots.

'We call this dressing the reed,' says Eric. 'It's really just cleaning it up but it's most important because if reed isn't cleaned after you've cut it you can't get it into neat bundles.' In an average season Eric will cut something between 4,000 and 6,000 bundles.

'We don't sell reed by the ton or the yard – we sell it by the bundle. That's the way it's always been done. A bundle used to be what the old boys decribed as "tied by a fathom". That meant you got five or six bunches of reed and when you stacked them upright you could tie them round with a piece of string measured at 6 feet long. For me these days a bundle is just what I can get my arm round.'

Norfolk reed is cut from mid-December until 5 April but that doesn't mean Eric is out cutting every day right through the winter; far from it, in fact.

'Well, the great difficulty is that reed needs water and lots of it, but if there's too much water you can't cut it because you can't get into the marsh! We have a system of sluices which can help keep the water levels where we want them, but if there's too much rain even the sluices won't help. We need a good easterly wind – that drives the water out.'

Most of Eric's thatch is sold locally despite the temptation of cheaper foreign imports. 'Well, I've got my favourite customers and I don't like to disappoint them!'

Norfolk reed varies in length between 4 feet and perhaps 6 or 7 feet and the best bit, as Eric explains, is the bit at the bottom. 'You see, that's the fat or butt end. It's the end that's always been in water so it's the strongest. It's the bit you will see when you look at a thatched roof. The higher, tapering part of each bit of reed will not be visible.'

Eric's marsh covers 365 acres on both sides of the River Ant about halfway between Yarmouth and Norwich. He lives where he works, too, and loves the fact that when he's at work he can always see his house in the distance.

In addition to reed he cuts sedge which is used for ridging work and both reed and sedge are cut on a two-year cycle.

'That's a lot better for the wildlife,' says Eric. 'Less disturbance you see, and there's a lot of wildlife here. I regularly see bitterns and otters for example. Both, I think, are making a bit of a comeback. Being able to work in a place where you see so much wildlife is part of the appeal.'

Once the reed is bundled up under Eric's arm he carries it out to a specially prepared area where dry chaff, or shucks, have been laid to provide a dry base.

'You never tie wet reed or stack it on wet ground – that's

the golden rule. We stack the bundles about ten high so that the wind can get into them and we come along later with the boat to take the whole lot off the marsh. Yes, it's a wonderful life. Keeps you fit and in touch with nature, but I don't think any of my children will take over from me – people now want a different life; not out here alone on the marsh.'

When he's not cutting reed Eric clears the ditches and gives talks to visitors and to local children about the way of life of the old marshmen. But he's pessimistic about the future:

'In ten years I think hand mowing reed will be a lost art. It will be gone, although I'd be surprised if the demand for reed ever completely died out.'

Page 146: Collecting the reed prior to cleaning it
Page 147: Cleaning the reeds with a special rake before bundling it

Left: Reed cutting can be lonely work, but Eric enjoys the peace and beauty of the area

Below: Reed is sold by the bundle, as it has been for centuries

The Arrowhead Maker

Hector Cole began forging iron at the age of four. As soon as he was able he turned what had been a childhood hobby into a career, and after more than fifty years he still finds the world of the forge utterly compelling.

'It's been my whole life really,' he says with a grin. 'What started as a little boy's fascination with swords ended up as a fascination with everything ancient and made of metal. But I'm not really a blacksmith in the generally accepted sense. Basically I make arrowheads and other iron items exactly as they would have been made centuries ago. In some cases I'm making items that haven't been produced for thousands of years.'

Inevitably these days many of the arrowheads, cannon barrels and other items made by Hector are commissioned by museums, but individuals sometimes order items that they plan to actually use.

'Well you see, there are lots of re-enactment societies,' explains Hector, 'and their members want things to be historically correct.'

Museums are in a slightly different position. Where once they expected visitors merely to look at artefacts from the past in glass cases, these days they are expected to be far more interactive.

'What it boils down to,' says Hector, ' is that these days visitors to museums want to handle these old objects, but you can't let thousands of people loose on a priceless, fragile sword from the Roman or Saxon period, or whatever. The solution is to commission me to make an exact replica of the item.'

Much of Hector's early fascination with old iron implements and weapons stems from his own interest in longbows. When he took up the sport he discovered that arrows of the type that would have been used when the longbow dominated warfare were not available, so he started making his own. Today before he makes an arrowhead, or any other item for that matter, he carries out exhaustive research into the relevant period.

Opposite: The arrow man: Hector Cole working the furnace in the forge

Below: Hector examines a medieval arrowhead. 'Most people think an arrowhead is a simple thing, but I have some twenty-two different types on my books'

Hammering the red-hot iron into shape

'Mind you, ancient methods – even methods used two thousand years ago – weren't all that different, at least in essentials, from those we use today. When you are working with iron you heat it until it can be forged, which means hammered, into shape. In the old days, of course, all the work was with pure iron. These days the vast bulk of metal work is with mild steel which is more tolerant of what we might call average workmanship. Wrought iron is much less tolerant – if you don't use it properly it is very unforgiving. It will split and bits will break off, but used in the right way it is a wonderful material.

'I started making arrowheads for other people when someone spotted a few that I'd made for myself. I made a batch for him and the thing took off from there. The arrowheads are forged from iron just as they were in medieval, Roman and prehistoric times, and to designs that were current in these different periods of history.

'Most people think an arrowhead is a simple thing, but I have some twenty-two different types on my books. There are hunting arrowheads and war arrowheads for example – both very different. An arrowhead designed for medieval war might be made specifically, for example, to pierce chain mail. This would be what we call a needle bodkin, and as the name suggests it is made long and very thin in order to get through the chain links of the mail. Needle bodkin arrowheads can be anything from 2 to 8in long.

'For hunting you need a head that cuts, so hunting arrowheads tend to be large and broad. Then there are fire heads – these are really like a small iron cage on the end of the arrow that was used to hold burning material. This might have been mutton fat and hemp or a gunpowder impregnated material. Then there are practice heads – simple short cones. There are also national differences – for example, I've just made an eighteenth-century Japanese arrowhead that has a cherry blossom shape cut into it.

'The biggest arrowhead I've ever made was 4in across and 5in long – originally something of that size would have been used for big game like deer.'

Describing the importance of his craft in earlier ages, Hector explains, 'You have to remember that when iron first came into Europe from the Middle East it was more precious than gold and the art of the metalworker was highly prized. Ironmasters created the weapons of war, and war was very important to our ancestors. The reason guns have barrels has everything to do with the ironmaster's skill, for example. Before mild steel and modern gunmaking techniques, gun barrels were made just as barrels for beer were and still are made. Hoops of red-hot metal were put round staves of metal, and when the hoops cooled they gripped the separate pieces tightly to make a blast-proof tube.'

Arrowheads in wrought iron are very expensive, explains Hector, because they are difficult to make and, perhaps more significantly, because if he makes a seventeenth-century arrowhead he uses seventeenth-century metal.

'Getting hold of iron of the right age to recycle is very difficult. I usually get it from old buildings – nails from beams, bars and bits of chain can all be reused. I keep a stock of iron from all periods from the twelfth century on, and as you can imagine it's very precious.

'After all these years I can usually tell just by the look and feel of a piece of metal if it is wrought iron rather than steel or whatever. The other thing to bear in mind is that you can no longer buy new wrought iron – it hasn't been made commercially in Britain since 1969.'

Hector is always busy – partly because he is the only maker of old arrowheads still working, and partly because the demand for them has risen. He has made arrowheads and other artefacts for institutions all over the world. In Britain customers have included Warwick Castle, the Royal Armouries and numerous individual collectors. Among Hector's more unusual commissions was one to make the gates for the Prince of Wales' house at Highgrove. He has also just completed a set of windows for an ancient house in Malmsbury. He is currently making a pair of shears for cloth cutting. Nothing exceptional about that, you might say, but the shears are 4ft long!

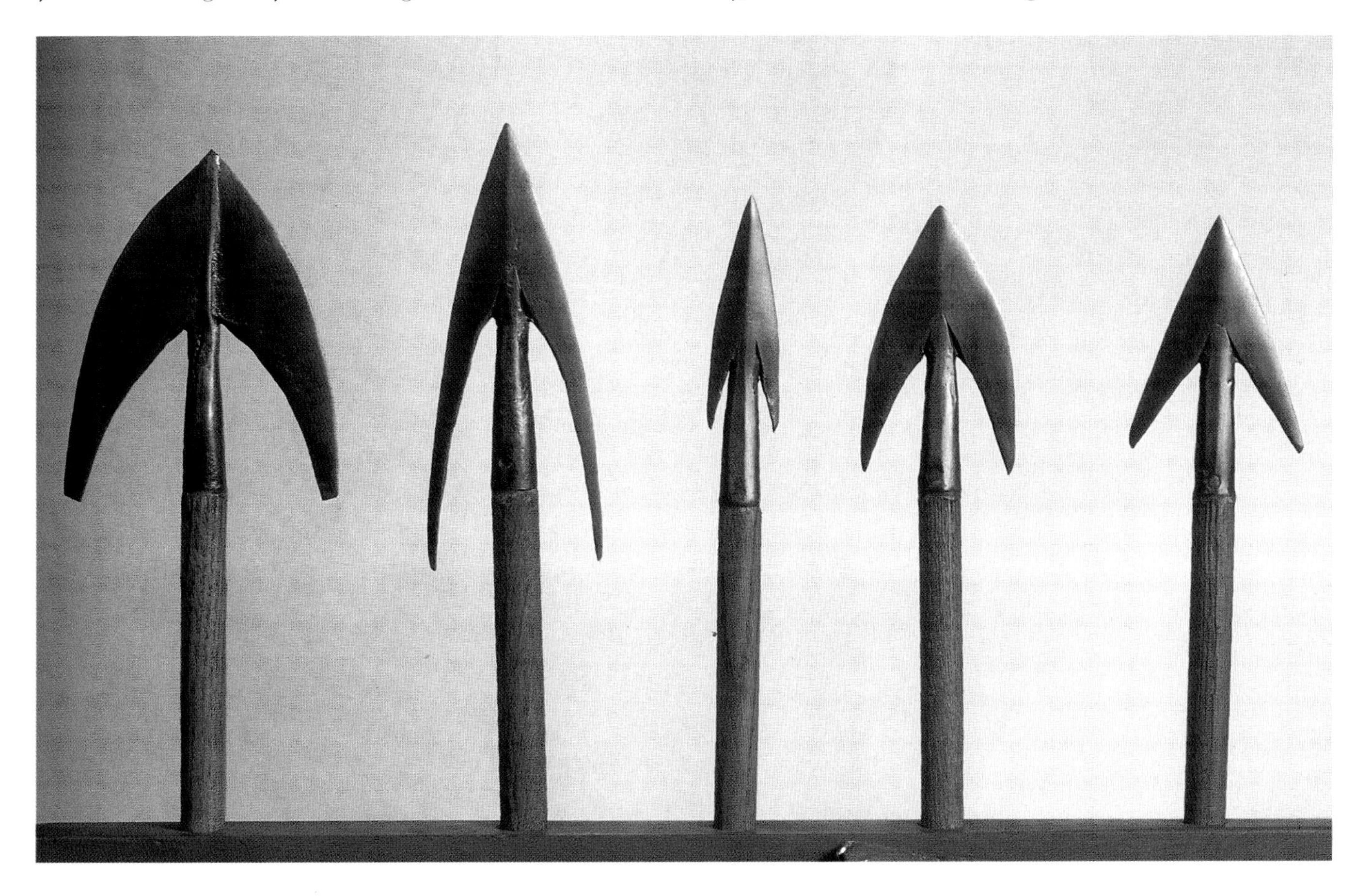

Hector produces a large range of arrowheads, all for different purposes

DEVON
EVEREST

The Golf Club Maker

No one really knows when golf was first played. The *Oxford English Dictionary*'s earliest reference to the game comes from a legal document of 1491 – a document which complains that golf is interfering with archery practice – but almost certainly people had been playing golf for a long time by then. Over the centuries the rules of the game and the equipment used to play it have changed out of all recognition, and today hi-tech materials produce clubs of enormous power and sophistication – but at least one family still makes golf clubs exactly as they were made centuries ago.

Three generations of the Davies family of Westward Ho! in Devon represent Britain's last traditional golf club makers. Eddie Davies, born in 1914, is the grand old man of the business: he started his apprenticeship in 1928 with the great club maker Charles Gibson of Musselburgh in Scotland. Says Eddie: 'Gibson learned from Dunne, another great name in golf club making, and Dunne was part of a tradition going back to about 1740.'

The techniques and materials used today by Eddie and his son John and grandson Sean are remarkable in that they are pretty much as they would have been in a golf club workshop in the 1700s. Eddie normally has two or three clubs on the go at any one time, but each involves a total of two full days' labour. For Eddie there is nothing nostalgic or sentimental about making golf clubs – producing these magnificent pieces of equipment is just his job.

Things were particularly tough in the 1960s and the traditional golf club virtually died out, but Eddie always believed that the market would revive, and that the desire to own a beautifully made golf club would eventually compete well with the desire for the latest soulless, if super-efficient, piece of equipment. And time has proved him right, because the three generations of Davieses now cannot keep pace with the orders that come in from all over the world. Eddie explains that they export most of their clubs to America where they are often given as prizes in golf tournaments, '... but they are also used by serious golfers with a taste for how it ought to be done,' says Eddie.

But how is a traditional golf club made? Eddie explains: 'It's actually quite simple. The trick is not in the tools or the design; it's in being able to do everything by eye – no measurements are taken. It's all done by what you might describe as instinct, and it takes many years of practice to build up that instinct. All the tools we use are pretty standard cabinet maker's tools – chisels, rasps and so on – and most of the ones I use today were the ones I started with in 1928.

'Beech is nearly always used for the head. It's probably been the preferred timber since about 1830. Before that, blackthorn was popular, and any fruitwood really – the most important thing is that the wood for the head has to have a nice tight grain so it can withstand the molten lead we pour in once it's been hollowed out. The lead gives

Opposite: The grand old man of wooden golf club making: Eddie Davies carries on a tradition going back to about 1740

Left: Rasping the head: Eddie at work on the head of a driver. Astonishingly he does all his work by eye – nothing is measured or marked out in advance

the club its power. Smoothing and finishing takes a lot of skill because the head of the club should be silky smooth with a wonderful shine.

'The club head is always fitted with a ramshorn tip – this used to be essential because in the days before modern, carefully tended golf courses, there were often stones and chips on the playing area, and the ramshorn prevented the wood being damaged.

'The shaft of a golf club has been hickory for as long as anyone can remember, and it's the perfect wood: pliable, strong and very straight. We use a spoke shave to get it to the right diameter and it is then spliced to the head. We don't use a socket joint because that wouldn't be strong enough; we use what's called a scare – a splice. After the splice has been glued, it's whipped up about 6in from the bottom with thick, tarred thread.'

In the Davies workshop even the putters are entirely of wood. 'We still make a long-nosed putter to an 1840s design,' explains Eddie; 'it has a head about 6in long. We also make a lot of miniature golf clubs, although the demand for full-size clubs is remarkably healthy. I think there will always be a demand for a club which has all the feel of golfing history. In fact, time seems hardly to have touched the craft.'

Addresses of Craftsmen

The Bark Tanner
Molly Arthur
The Old Manse
Grogport
Carradale
Argyll
PA28 6QL
01583 431255

The Bee-skep Man
David Chubb
Box Bush Farm
South Cerney
nr Cirencester
Gloucestershire
GL7 5UB
01285 860648

The Boot-tree Man
Bill Bird
Northwick Business
Centre
Blockley
Gloucestershire
01386 700855

The Cane Rod Builder
Shaun Linsley
Hand Made Split Cane
Rods
Jessamy
Mill Lane
Stour Provost
Gillingham
Dorset
SP8 5RA
01747 838251

The Cheese Maker
Charles Martell
Laurel Farm
Dymock
Gloucestershire
GL18 2DP

The Clay Pipe Maker
c/o Peter Ellis
The Manager
Wilsons & Co
Sharrow Mill
PO Box 32
Ecclesall Road
Sheffield
S11 8PL
0114 2662677

The Fan Maker
John Brooker
The Fan Attic
The Square
East Rudham
King's Lynn
Norfolk
PE31 8RB
01485 528303 (tel & fax)

The Horse-collar Maker
Terry Davis
5 Leamoor Common
Wistanstow
Craven Arms
Shropshire
SY7 8DN
01694 781206

The Kipper Smoker
Alan Robson
L. Robson & Son
Craster
Nr Alnwick
Northumberland
NE66 3TR
01665 576223 (tel)
01665 576044 (fax)

The Legal Wig Maker
c/o Jill Godfrey
Ede and Ravenscroft
93 Chancery Lane
London
WC2A 1DU
0171 405 3906

The Pole-lathe Man
Mike Abbott
Green Wood Cottage
Bishop's Frome
Worcestershire
WR6 5AS
01531 640005

The Pub Sign Painter
Mike Hawkes
Hazeldene
Whiteway
Stroud
Gloucestershire
GL6 7ER
01285 821292

Opposite: Leon Pezzack (p130) planing the blade of a traditional oar. Oars are made from several carefully prepared timbers, including elm for the tip of the blade, spruce for the handle and redwood for the main part of the blade

The Rake Maker
John Rudd
The Sawmill
Dufton
Appleby
Cumbria
CA16 6DF
01768 351880

The Rhubarb Forcer Maker
John Huggins
Ruardean Garden Pottery
West End
Ruardean
Forest of Dean
Gloucestershire
GL17 9TP
01594 543577

The Ships' Figurehead Carver
Steve Conway
Unit 1D
Grange Hill Workshops
Bratton Fleming
Devon
EX31 4UH
01598 710800 (tel & fax)

The Swill Basket Maker
Owen Jones
Spout Meadow
High Nibthwaite
Ulverston
Cumbria
LA12 8DF
01229 885664

The Brick Maker
Coleford Brick and Tile Company
The Royal Forest of Dean Brickworks
Cinderford
Gloucestershire
GL14 3JJ
01594 822160

The Thames Watermiller
Mildred Cookson
13 Littlestead Close
Reading
Berkshire
RG4 6UA
01189 478284
mildred@stonenut.demon.co.uk
or, for visitors to the mill and party bookings:
Mapledurham Water Mill
Mapledurham
nr Reading
Berkshire
RG4 7TR
01189 723350
01189 724016 (fax)
www.mapledurham.co.uk

The Somerset Hurdle Maker
James Hector
English Hurdle Company
Curload
Stoke St Gregory
Taunton
Somerset
TA3 6JD
01823 698418

The Lobster-pot Man
Nigel Legge
The Forge
Church Cove
The Lizard
Cornwall
TR12 7PQ
01326 290716

The Irish Bagpipe Maker
Eugene Lambe
The Pipe Maker
Kinvara
Co Galway
Ireland
00353 91638111

The Freeminer
Gerald Haynes
Wellmeadow
Lydney Road
Bream
Lydney
Gloucestershire
GL15 6EN
01594 563553

The Wooden Oar Maker
Leon Pezzack
Poldhu House
Fore Street
Mousehole
Cornwall
TR19 6TH
01736 731655

The Trug Maker
Contact Mrs S. Page
The Truggery
Cooper's Croft
Herstmonceux
Hailsham
Sussex
BN27 1QL
01323 832314

The Balloon Basket Maker
Darrell Hill
Somerset Willow Company
Wireworks Estate
Bristol Road
Bridgwater
Somerset
TA6 4AP
01278 424003

The Arrowhead Maker
Hector Cole
Couzens Farm
The Hill
Little Somerford
Wiltshire
SN15 5BQ
01666 825794

Index

Calligraphy by Susanne Haines

A DAVID & CHARLES BOOK

First published in the UK in 1999

A catalogue record for this book is available from the British Library.

ISBN 0 7153 0726 6

Book design by Tim Noel-Johnson
and printed in Hong Kong
by Hong Kong Graphic and Printing Limited
for David & Charles
Brunel House Newton Abbot Devon